BIOMAGNETIC HANDBOOK

Today's Introduction

to

The Energy Medicine of Tomorrow

by

William H. Philpott, M.D.

with Sharon Lindsay Taplin

Enviro-Tech Products, Publisher
17171 S.E. 29 Street
Choctaw, OK 73020
405/390-3499

ISBN #0-9636964-0-8

Biomagnetics is a natural science dealing with the application of magnetic fields to living things. This Handbook is written to provide basic information and understanding of biomagnetics as well as to guide individuals undergoing a medically supervised research project. Information herein has been compiled from many sources (see References) in addition to basic clinical observations made by this medical author.

Material presented is educational only and is not to be construed as medical advice, diagnosis or self-help prescription for any individual or any specific health problem. Any person experiencing a health problem should consult a physician. Unsupervised self-treatment can be hazardous in cases of serious illness.

Authors, publisher or distributors take no legal responsibility whatsoever for the reader's use of contents herein. This book is intended only as a chronicle of scientific developments and research of magnetic processes and products.

INTRODUCTION

Magnetic Energy:
From Ancient Mystery to Space-Age Scientific Breakthrough

Magnetic energy, once an ancient mystery, is now recognized as one of the strongest natural forces in the universe. Magnetism is the basis for present day discoveries being made in all fields of applied sciences--electronics, electromagnetics, chemistry, physics, etc. From an industrial standpoint, magnetic energy is very familiar to everyone--although it may routinely be taken for granted. It plays a role in a wide range of practical everyday functions from simple domestic gadgetry to sophisticated navigational systems for jets, oceanliners, etc. The list of industrial applications could go on and on.

However, while focusing on the development of magnetics industrially, researchers have sadly neglected to study the effect of magnetic energy on life and health--until recent years. But now magnetic therapy, practiced and described by Egyptian and Greek physicians thousands of years ago, is no longer ancient history--or a mysterious force. It has been established by scientists that this invisible energy passes through all material matter affecting every living system on the planet. While magnetic fields surround the earth protecting living organisms from harmful solar radiation and emissions, they also influence behavior, mental function, physical energy and the biological welfare of man in general.

As biomagnetics unravels, the findings are astounding. Modern advanced medicine is now recognizing the potential value of using magnetic energy in treating chronic physical and mental disorders, tumors, etc. In May, l989, the Food and Drug Administration (FDA) and the University of Tulsa jointly sponsored the First Conference on Emerging Electromagnetic Medical Technology. Researchers, physicians and medical specialists from around the world gathered and shared information on therapeutic benefits of magnetic therapy for various medical disorders.

World governments, performing extensive research for space exploits, etc., have unveiled vital information on the human response to magnetic fields and magnetic pulsing frequencies finding that specific magnetic fields and specific pulsing frequencies evoke specific biological responses. For example, the Earth's natural pulsing frequency of 7.96 cycles per second reduces stress and encourages relaxation and sleep. In turn, sound restful sleep (a negative magnetic energy state) is likely the greatest producer of physical and mental energy. Such research findings introduce the early beginnings of Space Age Energy Medicine utilizing biomagnetic technology.

Medical Magnetics

For the past two decades the medical establishment has been focusing on biochemistry for answers in medicine. Biophysics has been sadly neglected but now appears to be on the threshhold of making equally valuable contributions to the scientific practice of medicine. It is evident that, in fact, biophysics actually controls biochemistry--the biochemical processes of the body being magnetically governed.

More recently discoveries in the field of biomagnetics have led to the development of many new diagnostic and therapeutic tools: Superconducting Quantum Interference Detector (SQUID) for determining minute magnetic fields including those in living organisms; magnetoencephalogram, an improvement over the electroencephalogram; nuclear magnetic resonance (MRI) for imaging which is replacing x-ray and alleviating the risk of cellular damage from radiation; Therafield and Diapulse electro- magnetic instruments for pain and edema; Xomed magnetic hearing aid for conduction deafness and the Cadwell Magneto-electric Stimulator. Diagnosis is enhanced by magnetic instruments which can detect areas of disordered body metabolism. And the list keeps growing.

The Journal of the American Medical Association in July l989 published an article, "Magnetism: A New Method for Stimulation of Nerve and Brain" which referenced thirty-three articles in medical journals in the last two years on the use of magnetic energy in nerve testing.

Even with such advancements and acceptance of magnetic instruments, for unexplained reasons resistance continues to linger in medical circles concerning the use of magnetic energy in general treatment of biological systems. It is time that we all know the facts about magnetic energy as it applies to biological systems and begin to apply that knowledge more vigorously.

Many scientists have contributed to our present state of knowledge about biomagnetism. Robert O. Becker, M.D., has observed that there appears to be one fundamental force that heals--electromagnetic energy. Andrew Marino, biophyscist, has stated that a new biology must be developed in which electromagnetic energy receives the critical consideration and evaluation that it merits on the basis of present knowledge.

It has been supported, through multiple animal experiments, that there are distinct and opposite effects of the two separate magnetic poles on metabolic function. In terms of biological response, the separate negative and positive

polarities are as distinctly opposite as day and night, hot and cold and acid and alkali.

Effects of
Magnetic Energy on Living Metabolic Systems

Compiled from other's research as well as the clinical observations of this medical author:

Negative Magnetic Field Energy	Positive Magnetic Field Energy
Increases cellular oxygen*	Decreases cellular oxygen*
Pulls fluids & gasses	Pushes fluids & gasses
Reduces fluid retention	Increases intracellular edema
Encourages deep restorative sleep	Stimulates wakefulness
Fights infection	Accelerates microorganism growth
Supports biological healing	Inhibits biological healing
Reduces inflammation	Can increase inflammation
Normalizes acid base balance	Acidic metabolic response
Relieves/stops pain	Increases pain
Relieves withdrawal symptoms by reducing inflammation (nonaddicting)	Relieves withdrawal by endorphin production (addicting)
Can reduce/stop symptoms	Can intensify symptoms
Promotes mental acuity & reasonableness	Promotes mental overactivity & unreasonableness
Promotes bright happy affect	Promotes dull depressed affect
Reduces/dissolves fatty deposits	Encourages fatty depositing

*Indirect evidence

Note: The negative magnetic field is equal to the Earth's north pole, negative ionization and negative DC circuit while the positive magnetic field is equal to the Earth's south pole, positive ionization and the positive pole of a DC circuit.

Characteristically negative magnetic energy normalizes and calms while positive magnetic energy disorders and overstimulates the biological system. Microorganisms and parasites (including those harbored in the human) overgrow with positive magnetic energy exposure and are inhibited by negative magnetic pole exposure. The heart can be stimulated to tachycardia (harmful stress

effect) by direct prolonged exposure to positive magnetic energy and slowed slightly by negative magnetic energy (valuable antistress effect).

Since biological response to magnetic energy does exist, magnetic energy can be used in both diagnosis and treatment of health problems. However, to therapeutically apply magnetic energy safely and effectively it is necessary to separate negative magnetic energy from positive magnetic energy (unipole magnet) so that the energy from either pole can be applied separately as indicated by the condition.

The brain and spinal cord of the body are positive pole oriented in relationship to body tissues peripheral to the central nervous system (CNS) which are negative pole oriented. Most human illnesses (emotional disorders, seizures, infections, toxic states and acidic states) are positive magnetic energy driven. It has been clinically observed that positive magnetic over-stimulation of the CNS and strong positive magnetic disorders in peripheral tissues can be materially corrected by local application of negative magnetic energy. However, the negative magnetic energy exposure must be of sufficient intensity and duration to counteract the abnormal positive magnetic energy thus reestablishing the negative-positive magnetic energy balance of the tissues. The biological system can then effectively proceed with normalized healthy metabolic processes. When CNS positive magnetic energy overstimulation produces anxieties, phobias, obsessions, compulsions, depression, psychosis or seizures, the correct amount of appropriately applied negative magnetic energy can control the positive magnetic energy overdrive freeing brain function to normalize and proceed with normal perception and learning, including corrective programming of the brain.

The healing process, whether involving a cut, burn, blood clot, allergic reaction, addictive state, infection, toxic state, emotional or mental reaction or any other metabolic disorder, marches to the response of metabolically produced magnetic energy. Therefore, it behooves physicians and all who seek healing to understand the biophysical role of magnetic energy in biological systems. We now understand that the healing process need not be left to chance alone that a person's internal metabolic energy system will produce the necessary healing magnetic energy, but rather we can externally apply negative magnetic energy to the specific area of the body in need of healing.

THE BODY ELECTROMAGNETIC

All life--plant, animal and human--exists in a magnetic field, the Earth's, and all life responds to the magnetic field of the Earth. The Earth's magnetism activates an enzyme system in fruits and vegetables which causes normal ripening (demonstratable by placing fruit or vegetable on a magnet). Most animals exhibit a "homing instinct" while migratory birds fly south in the winter and homing pigeons are magnetically guided back to their birthplace. Researchers have proven animals respond to and are influenced by changes in the Earth's magnetic fields. What about humans?

Physical & Mental Activity: An Electromagnetic Function

The human body itself is an electromagnetic machine. Each body cell has a positive and negative field in the DNA. The cells are paramagnetic and assume the polarity of the magnetic field in which they exist. Further, tissue groups such as the central nervous system and peripheral tissues take magnetic polarities opposite to each other. All physical and mental energy and function are controlled by chemical electromagnetic impulses to and through the brain and central nervous system.

Pineal Gland: A Magnetic Organ

The pineal gland, a pinecone-shaped structure in the exact geometric center of the head, is a magnetic organ containing magnetite crystals. This gland is responsible for producing melatonin which is the master controller of hormones, enzymes, immune function, oxidation, carbohydrate metabolism, pigmentation by melanin and the cyclic pattern of sleep-wakefulness. This master gland governs 1) stress levels and reactions and 2) antioxidant production for the control of harmful free oxidative radicals. It has an anti-aging effect.

Since the pineal gland is magnetic, it is sensitive to the daily cyclic patterns of the Earth's magnetic field and its function can be affected by the application of magnetic field exposure. As Robert Becker states in Cross Currents, "Melatonin secretion in human subjects may be changed at will by exposure to steady magnetic fields of the same strength as the geomagnetic field". When melatonin is deficient energy expression is disorganized resulting in diseases.

Biochemistry: An Electrochemical Function

As previously mentioned, biochemistry is actually controlled by biophysics--and even the nutritional building blocks such as amino acids require the energy of a

magnetic field for formation.

Cell Division: A Magnetic Process

The body itself is composed of a large number of individual cells which, depending on their identity, group together forming specific tissues and organs. Tissues and organs are distinguishable by their individual cellular function--nerves, blood, bones, muscles, brain, liver, etc. Cells are alive, constantly forming and renewing themselves in a process called cell division, the process responsible for keeping the body healthy and revitalized. The force which activates the formation and division of cells is magnetic energy.

Not only are individual cells dependent on magnetic energy for proper function, but the interaction of all the organs and systems in the body is dependent on the constant presence and influence of magnetic energy. Of equal importance to magnetic fields are vibrational frequencies which evoke specific biological responses. The body's natural vibrational frequencies are controlled by the balance between positive and negative magnetic fields with an increase in positive magnetic energy increasing frequencies and an increase in negative magnetic energy decreasing frequencies.

If and when a magnetic imbalance and/or oxygen deficit occurs in an organ or system, function is disturbed producing symptoms and the condition is diagnosed as an "illness". For example, when there is insufficient oxygenation and/or negative magnetic energy present for normal cell division (normal healing process), there is a high risk for cell mutation with rapid cellular division. The best treatment is the application of negative magnetic therapy which pulls oxygen into the affected area providing l) negative magnetic energy and 2) oxygenation of cells unlocking the abnormal growth cycle and allowing normal cell division to proceed.

Considering that man truly is an electromagnetic creature with every cell in the living system an electrical battery (each cell having a negative and positive magnetic field in the DNA) and the function of the organs, glands, etc., electrochemical in operation, it becomes obvious that disorders and diseases are the result of electrochemical breakdowns which can be positively affected with appropriate application of magnetic energy.

MAGNETIC ENERGY: THE ENERGY OF LIFE

Energy is essential to everyone with no respect of persons--the poor as well as the rich, the old as well as the young, and the sedentary as well as the athletic. And the list could go on and on. . . . . Where does energy come from and where does energy go? In order to fully answer these questions, it must first be understood that human physical and mental energy are in reality magnetic energy which is produced in several different ways.

SOURCES OF BIOLOGICAL ENERGY

Biochemical Production of Energy

One of the major sources of physical and mental energy is, of course, the biochemical processing of food and water into energy. As food is eaten, salivation begins the chemical breakdown of the food, digestive processes continue, nutrients are absorbed into the blood, cells, nerves, etc., ultimately producing electromagnetic energy--the energy of life. Life involves a multitude of physical and mental responses all governed by electromagnetic energy--sight, hearing, movement, sensation, metabolic processes, cell health, emotions, thoughts, etc. Therefore, from the biochemical standpoint, physical and mental function are dependent on efficient l) digestion, 2) absorption and 3) chemical processing of foods and nutrients.

Sleep: Nature's Energy Recharging Process

Sleep is Nature's own recharging process--which restores energy in the biological system. Optimum physical and mental energy actually consist of a delicate balance between excitation (wakefulness & activity) and inhibition (relaxation & sleep). States of excitation cannot be maintained for extended periods of time without undue stress and fatigue leading to dire metabolic consequences. Organ functions and metabolic processes can become over- stimulated, overworked and prematurely wear out. Obviously it is unhealthy and risky to stimulate energy with the use of caffeine, amphetamines, etc., rather than restoring energy with proper rest and sleep.

During the waking hours the positive magnetic energy drive of physical activity and mental concentration is higher than the negative magnetic energy drive and "burns up" energy. During sleep the negative magnetic energy drive is higher than the positive magnetic energy drive in the body and:

l) restores physical and mental energy
2) trains out anxieties through the dreaming process

3) promotes metabolic processing of toxins
4) prevents &/or reverses infectious states
5) supports biological healing processes
6) stimulates production of melatonin by the pineal gland

Upon awakening from a restorative sleep, negative and positive magnetic energies are balanced (equal) so appropriate physical and emotional responses can be initiated (positive drive) or inhibited (negative drive) at will and adverse environmental influences can be counteracted.

Absorption of Magnetic Energy

In addition, a degree of human energy is l) absorbed from the earth (surrounding magnetic field) and also 2) inhaled via oxygen (paramagnetic) which carrys the earth's magnetic field into the human cell. The higher the surrounding magnetic energy, the higher the human level of energy and function.

The Smithsonian Institute has demonstrated that sharks obtain a large degree of their energy from space (magnetic fields) rather than food. Insects obtain 90% of their energy from the earth's magnetic fields. Interestingly, sharks and insects are both highly resistant to viral infectious diseases including cancer. Insects do not have an immune system. Negative magnetic field energy is alkalizing (pH normalizing), anti-inflammatory and oxygenating, all of which can solve many problems without involving the immune defense. It would appear probable that the first line of defense in health is actually negative magnetic field energy with the immune response being the second line of defense.

Other Sources of Energy

It is a law of physics that the flow of fluid through a magnetic field produces an electromotive force. For example, Niagara Falls, a natural occurring phenomenon, produces electromagnetic energy from the water flow over the falls and through the earth's magnetic field. Similarly, it can be reasonably assumed that the flow of blood through the body (in the earth's magnetic field) produces an internal source of magnetic field energy.

A degree of energy (magnetic) is also derived through bone stress. Bones, piezoelectric, produce electromagnetic energy from flexion and stress during exercise. In addition, electrolysis of cerebral spinal fluid during sleep also produces biological energy.

So it becomes obvious that no matter what the source, physical and mental

energy are ultimately electromagnetic energy--which can be measured by the sophsticated SQUID instrument which monitors the magnetic energy emanating from a living being. And when death occurs the SQUID no longer registers magnetic energy in soft tissue of the body--the magnetic energy which produced physical and mental function no longer exists!

Life's energy is not a single energy but is a balance of two distinctly separate energies--negative magnetic energy and positive magnetic energy. Humans are electromagnetic organisms plugged into the magnetic fields of the earth, sun, moon, etc., and are influenced by external exposures and changes as well as internal stressors.

As can be seen, it is very important to be aware of the sources of biological magnetic energy since there is the capacity to alter these factors for the better.

WHERE DOES ENERGY GO?

Biological energy is "burned up" in daily pursuits, "sapped" by internal stressors and "wiped out" by assaulting external exposures. Through extensive research in biomagnetics it has been established that a proper BALANCE between negative and positive pole magnetic fields in the human organism is necessary for life, health and happiness. It has also been observed that the two magnetic poles have distinctly opposite effects on biological systems. Thus, the BALANCE between negative and positive magnetic field energy is critical because positive magnetic energy stimulates and with prolonged exposure disturbs human metabolism while negative magnetic energy normalizes human metabolism. (See Effects of Magnetic Energy on Living Metabolic Systems.) When positive magnetic energy strength exceeds negative magnetic energy strength for very long physical &/or emotional symptoms occur. In simple terms, health is a balance of negative/positive magnetic energy while all conditions opposing health are positive magnetic energy expressions.

Actually health is a battle of energies--negative/positive magnetic energy balance vs. positive magnetic energy attackers (internal & external). Internal attackers include allergens (foods, animals, plants, etc.) nutritional deficiencies, addictions (caffeine, nicotine, drugs, alcohol, etc.), infections, metabolic disorders, etc.

External influences on human magnetic energy balance include chemical toxins (agricultural, printing, colognes, auto exhaust, petrochemical hydrocarbon products, etc.), electromagnetic contamination and a possible deficiency in the earth's magnetic field today.

Thus we can easily see the formula for low energy, poor function, multiple symptoms and disease:

Internal Stressors
Allergies/Addictions
Nutritional deficiencies
Infections/Toxins
Metabolic Disorders
Scar Tissue

= Magnetic Energy Imbalance = Physical &/or Emotional Diseases

External Influences
Electromagnetic Contamination
Magnetic Deficiency in Earth
Geomagnetic Disturbances
Chemical Exposure/Toxins
Allergens (Pollens, molds, etc)

With sufficent negative magnetic energy, the battle for health can be won. When the body is overstressed with positive magnetic energy attackers, external negative magnetic energy support can be supplied with unipole magnets.

Effect of Geomagnetic Disturbances
on Biological Function

For a number of years scientists have been aware of the constantly changing magnetic fields on the earth and have been studying the impact on human health and behavior. Local magnetic fields are altered during thunderstorms while fluctuating magnetic fields from outer space (sun spots, solar flares, moon phases, etc.) episodically jar the earth. When sun flares jar the earth's magnetic field they produce a higher ratio of positive magnetic energy to negative magnetic energy. Such invisible subtle changes in the geomagnetic field affect the human nervous system by altering the body's electromagnetic field and people develop symptoms. Law enforcement, psychiatric and emergency room workers will testify to the increased numbers of people who become emotionally disturbed and require hospitalization during such times. In addition, researchers have noted an increase in heart attack rates during periods of geomagnetic disturbance. The population so affected either physically or mentally at such times obviously was at high risk (in a precarious negative-positive balance) preexisting the incident. Magnetic therapy as described herein has shown to be effective in stabilizing many patients in such precarious health states.

In May l989, a splendid display of "northern lights" flashed across western U.S.

skies. Coinciding with the natural phenomena were massive disruptions of radar, satellite, navigational and radio communication systems. Power surges in oil pipelines, telephone lines, undersea cables and electric systems occurred. Electrical blackouts occurred in some areas of the world. The spectacular display and resulting disruptions were the result of solar storms raging millions of miles away on the sun. Sunspots erupting on the surface of the sun boil forth magnetic energy, hot gasses and radiation causing solar flares, each flare capable of unleashing the energy of 10 million hydrogen bombs.

Of course, to some degree the surrounding atmosphere protects the earth from the impact of solar storms--but not completely. It has been observed that in addition to disrupting communication systems, etc., the invisible magnetic fluctuations and pulsations from such solar disturbances influence certain animal species. For example, homing pigeons have been observed to suffer navigational difficulties during such geomagnetic disturbances while animals often display restlessness and unease.

Does it seem so strange then that mankind, a very magnetic being, would be directly affected by such disturbances in the earth's geomagnetic field? Not only have U.S. researchers long associated disturbances in the earth's geo- magnetic field with an increase in psychiatric admissions and behavioral changes, but laboratory tests have illustrated that magnetic fields can interfere with cellular DNA, the blueprint of life itself, with the possibility of mutations directly arising from short-term changes in the earth's magnetic field.

Effects of Electromagnetic Contamination

Another new environmental health threat appears to be looming on the horizon--the threat of chronic low-level electromagnetic radiation from power lines, electric blankets, household appliances, computer terminals, etc. Nancy Wertheimer, Ph.D., scientist in Boulder, CO, along with Ed Leeper, Ph.D., published research findings in the American Journal of Epidemiology concluding that children living in homes near overhead electrical wires carrying high current died of cancer at twice the rate as children living in homes near low-current wiring.

Household appliances such as hair dryers, toasters, etc., can also produce strong magnetic fields and pulsing frequencies but do not constitute continuous exposure as they are used only episodically. On the other hand, electric blankets and electric dial-face clocks offer chronic low-grade magnetic exposure. Other contamination sources include TV and VDT (video display terminals) which often produce strong magnetic fields up to 1-2 feet from the screen.

Keep in mind that anytime an electric current is present there will be an associated negative-positive magnetic field. Prolonged simultaneous exposure to a low (yet beyond human adaptation) or high gauss strength of both magnetic fields has been observed to be capable of producing positive magnetic field effects (lethargy, fatigue, illness, etc.) on biological systems. Thus prolonged electromagnetic exposure can upset the magnetic balance (negative-positive) of the body disordering the body's energy systems including brain function.

Additionally, remember that the cells, organs and entire body have a unique functional vibratory rate (See Body Magnetic and Electric) and react to subtle energy flows. Consider the 24-hour-a-day exposure of the body to frequencies and electrical waves--not only from 60 cycle AC wired into the home and workplace--but the invisible waves from radio, telephone, police, fire and utility signals--and the microwaves--all with varying vibratory rates constantly bombarding the body while awake or asleep. All these environmental electric pulsing frequencies differ from those of human adaptation interfering with metabolic functions.

Bear in mind that the natural pulsing frequency of the Earth is approximately 7.96 cycles per second (CPS) which is the healthy vibrational frequency of the human body as a whole. When asleep the rhythm of the brain runs from 1 to 2 CPS, when awake and alert 8-12 CPS and with mental concentration increases to 18-22 CPS. Thus it can be seen that 60 cycle AC electric current and the varying frequencies of radio waves, etc., are disordering to the biological metabolic function of animals and mankind. In today's highly industrialized world, humans reside in a virtual sea of metabolically disturbing electro- magnetic frequencies and biologically disordering positive magnetic fields. What can be done to protect against electromagnetic contamination and preserve energy and health?

Number one, negative magnetic field therapy as set forth in this handbook can provide protection against harmful magnetic fields and also can be of therapeutic value for existing physical and emotional illnesses where the magnetic balance has already been disrupted. Refer to Electromagnetic Interference in the alphabetical listing of disorders in this Handbook.

Other advisable precautions include:
- Sit at least 2.5 to 3 feet from any TV or video screens.
- Discontinue use of electric blankets until redesigned to reduce magnetic field emissions.
- Unplug water bed heaters when occupying bed.
- Place dial-face electric clocks at least 3 feet from bedside.
- Investigate yard, home and immediate surroundings for possible electromag-

netic contamination areas & consider measure to reduce exposure.

Magnetic Field Deficiency Syndrome

Since a degree of human energy appears to be absorbed from the earth's magnetic field, what would happen if a magnetic field deficiency occured in the earth? Kyoichi Nakagawa, MD, Director Isuzu Hospital, Tokyo, Japan has spent 20+ years researching magnetism and humans. He concludes a magnetic field deficiency syndrome already exists which produces stiffness of shoulders, back and neck, chest pains, headache, heaviness of head, dizziness, insomnia, habitual constipation and general lassitude. Nakagawa references authorities demonstrating the earth's magnetic field has diminished 50% over the last 500 years. Is the human body adapted to a higher gauss strength and with the earth's weakening magnetic gauss strength human deficiencies are emerging? It would seem logical since when a patient with symptoms is exposed to a higher gauss magnetic strength, symptoms are considerably relieved or disappear.

Nakagawa points out that metal buildings, automobiles, trains, etc., are absorbing the earth's magnetic field causing interference and loss of gauss strength. Not only can the metal structures absorb the earth's magnetic energy but also can shield living beings from the earth's natural energy source. Consider that man often spends his waking and sleeping hours in metal vehicles, steel supported dwellings and steel frame superstructures of the work place--which may be having a deleterious impact on physical and mental energy levels. Obviously, such environmental conditions can interfere with the human energy system since eletromagnetic absorption can not occur at an optimum rate. Thus symptoms of magnetic field deficiency syndrome could emerge.

For symptoms of Magnetic Field Deficiency Syndrome refer to "Magnetic Field Deficiency Syndrome" in the alphabetical compilation of disorders in this Handbook.

Electromagnetic Interference From Amalgam Dental Fillings

Dental amalgam fillings set up a readily measurable direct flow of current between fillings with associated positive and negative magnetic poles. The DC (+) positive magnetic pole encourages microorganism replication. In addition, the magnetic field can provide sufficient positive energy to interfere with biological energy in general and brain function in particular. Weakness and ringing in the ears have been observed as primary symptoms of such electromagnetic interference.

Healthy persons with a full quota of negative magnetic energy are not likely to notice symptoms from amalgam fillings. However, as stress factors accumulate and positive magnetic field energy increases, the presence of the fillings can become symptom producing. In some cases raising (recharging) negative magnetic energy by use of magnetic therapy and sound sleep is adequate for biological energy needs. In cases which do not respond, however, replacement of amalgam fillings with nonmetal fillings may be necessary along with correction of other energy reducing stress factors.

Effect of Chemical Exposures on Human Function

In addition to the afore discussed energy "sappers", the inability to biochemically process toxins from chemical exposures can completely "wipe out" biological energy. Shopping fatigue, often blamed on "pounding the concrete all day", is more frequently the result of exposure to chemical toxins (petrochemical hydrocarbons) which exist in the store place and mall. Physical and mental energy are exhausted when the body is so overwhelmed with toxic exposure that the internal detoxificiation system cannot function fast enough to throw off the effects of the chemical assault. Chemicals in the mall which affect physical and mental energy include perfumes, fabric dyes and treatments, synthetic carpet materials and treatments, insecticides, plastic fabrics, building materials, paint, detergents, cleaning supplies, cigarette smoke, etc.

A 35 year old elementary school teacher, was so extremely sensitive to environmental substances that she had to quit teaching in the middle of the school year. Even chalk dust would cause such a violent reaction that energy would be completely wiped out and she would dissolve into a weeping nonfunctional state. Obviously, not only could she not work but she was quite unable to do even the most necessary shopping for survival. She had to hire everything done. Then magnetic therapy was tried. One can only imagine her delight when she found that if she wore a plastiform magnetic strip on her head under a hat that she could go shopping--and spend several hours in a mall without development of symptoms.

If chemical exposures from common everyday exposures in the store or schoolroom can have such an impact on human function and energy supply, imagine the chemically-created physical stress of anyone who works daily with petrochemical hydrocarbons (farmers, painters, builders, gas/oil laborers, plastic manufacturers, etc.). For information on magnetic therapy for chemical exposure refer to Shopping Fatigue and Chemical Exposures in the alphabetical listing of Disorders in this Handbook.

Maintaining, Restoring & Increasing Biological Energy

What can be done to preserve and increase energy? Proper sleep, application of negative magnetic field energy, exercise and proper diet and nutrition are the key ingredients in maintaining optimum energy.

- Negative magnetic energy: combats internal stressors and external influences, relieves stress, promotes sleep, promotes detoxification and oxygenates tissues naturally increasing energy levels
- Sleep: restores energy & rejuvenates the biological system
- Proper diet/nutrition: biochemically produces energy
- Exercise: bone flexion produces energy

The remainder of this book will be addressed to the health-promoting effects and applications of negative magnetic field energy. For further information on diet, nutrition, food allergies, etc., refer to Brain Allergies: the PsychoNutrient Connection and/or Victory Over Diabetes: a BioEcologic Triumph.

With sufficient negative magnetic energy the battle for energy and health can be won. When the body is overstressed with positive magnetic energy attackers, externally applied negative magnetic energy will provide support. Magnetic application should be considered not only as a preventive therapy but also as a therapeutic modality for symptom relief and/or illnesses.

MAGNETS

How long have magnets been around? Nature itself manufactured the first magnet ever--lodestone--a mixture of lava rock, iron and minerals. As molten lava spewed from erupting volcanos cooling and hardening it absorbed, and due to its iron content, retained the Earth's magnetism. With passage of time and the natural eruptions of the Earth the lava eventually broke up into small rocks--natural magnets.

Early man soon discovered that the lodestone was attracted to other stones that contained iron. And research of magnetic energy began! The ancient Egyptians discovered early on a great deal about magnetism and utilized its energy and power in pyramid design and construction.

Today a variety of man-made magnets are available--neodymium, ceramic, iron, plastiform--in a wide range of sizes and strengths. While natural magnets (lodestone) are weak, commerical magnets can be manufactured in different strengths for different purposes. Magnetic strength, power, energy is measured in "gauss" units.

All magnets have two poles--north pole and south pole. Electrons (magnetic energy) from the north pole spin left (counterclockwise) generating negative magnetic energy while electrons from the south pole spin right (clockwise) generating positive magnetic energy. This discovery has been reported and confirmed by sophisticated space-age instruments which measure magnetic energy.

In this publication north pole magnetic energy will be referred to as negative (-) magnetic field energy and south pole energy will be referred to as positive (+) magnetic field energy. The terminology "negative" and "positive" has no reference to "good" and "bad" but rather relates to the laws of electricity and physics.

Again, remember that negative and positive polarities have separate and opposite effects on biological systems! To review the differences, refer to "Effects of Magnetic Energy on Living Metabolic Systems".

Polarization

Not all magnets are created equal and persons working with magnets must be aware of the differences. There are unipole magnets and bipolar magnets. Unipole magnets are polarized with negative energy on one side of the magnet and positive energy on the opposite side. Researchers and health professionals

applying magnetic energy to biological systems must always use unipole magnets so negative or positive energy can be applied selectively.

When both positive and negative energy are present on the same side of a magnet, the magnet is referred to as bipolar. Bipolar magnets, popular in industrial use, are not to be applied to biological systems since both poles are present and cannot be separated. For example, it would be unwise to grab the decorative magnet from the refrigerator door and use it for therapy since it is bipolar to enhance "sticking quality". If bipolar magnets are applied to the human body, positive pole energy would be absorbed along with negative. Horseshoe magnets are not suitable for biomagnetics either since the two poles are physically too close together to allow either pole to be applied individually.

Remember, since negative and positive magnetic energies have opposite effects on biological systems, only UNIPOLE magnets are appropriate for application in biomagnetic therapy described in this Handbook.

Differences in Polarity Identification

Another warning! In purchasing magnets, the reader must be aware and cautioned that the poles of magnets are identified by two opposite systems: l) traditional method of identification used in industry and navigation (opposite of the earth's polarity) and 2) physics identification of the magnetic poles corresponding with the earth's magnetic poles. Biophysics recognizes and equates cellular and tissue polarity with the earth's magnetic poles.

Another caution, some authors (especially foreign authors) and medical practitioners have continued to use the traditional industrial method of identifying polarities. It is, therefore, of utmost importance before getting involved in biomagnetics that th Reader understand The Law of Polarity and establish the true identity of the poles (corresponding to earth's poles) of the magnets to be used.

The Law of Polarity: Opposites attract while like forces repel.

Methods for Correctly Identifying Magnetic Polarities

Compass method: When the compass needle (north-seeking) marked N points to the magnet it indicates the negative (north) pole side of the magnet (opposites attract).

Magnetometer method: This instrument will indicate by dial and needle

whether the magnetic pole is north pole (-) or south pole (+).

Magnet method: Use a correctly identified and marked magnet to determine the poles of an unmarked magnet. Simply try stacking the magnets and remember The Law of Polarity--opposite polarities attract and like forces repel.

Another simple method to determine polarity is to suspend a unipole magnet on edge from a string. Keep in mind, the Earth is a magnet with a North Pole and South Pole. The south (+) pole of the magnet will be attracted to and face the North Pole of the Earth. No matter how many times the magnet is spun, it will always reposition itself in alignment with the Earth's poles--opposites attracting.

Magnets Used in Research Studies

Following is a description of the permanent unipoled l) ceramic magnets and 2) plastiform magnetic strips used in biomagnetic research therapy in this Handbook.

- Magnetic bed (mattress pad) containing 850 or 4000 gauss unipole ceramic magnets providing negative pole exposure enhanced by vibratory rates

Magnet Size	Gauss Strength Rating 1	Rating 2
• 4X6X1/2" ceramic magnet	3950	900-1000
• 2X5X1/2" ceramic magnet	3950	900-1000
• 2 X 1 3/8" (cube) ceramic magnet	4150	1250-1350
• Neodymium Magnet (round) l" diameter X 1/2" thick	12,300	3500
• 2" plastiform magnetic strip	525	150-175
• 3" plastiform magnetic strip	2200-2450	250-275

NOTE: Rating l: Manufacturer's rating of residual gauss strength in a closed circuit. This is the preferred and most reliable method of identifying gauss strength. Rating 2: Electronic gaussmeter reading (hall effect) of surface strength in an open circuit.

The depth of penetration of magnetic energy is in direct relationship to the physical size (mass) of the magnet--the larger the magnet the deeper the penetration. For example, the small round neodymium magnet penetrates 2-3" into the body whereas the 4X6X1/2" magnet penetrates completely through the body of an average size person. The 4X6X1/2" ceramic magnet has much greater penetration capacity due to its physical size.

Magnetic therapy can take place l) while sleeping (with a magnetic bed), 2) in a reclining position with magnets under or over the affected areas or 3) while working, shopping, etc. In the latter instance magnets can be held in position with tape, sweatbands, Ace Bandages (3" for neck/head, 4" for back, thymus, chest, neck, abdominal areas), abdominal binders, hats, caps, or custom-sewn aprons with pockets sewn in position over the various organ systems.

How Magnetic Energy Works

In addition to the scientific findings by other scientists discussed herein, the medical author has observed and/or deduced the following regarding magnetic energy:

An initially nonmagnetized paramagnetic substance exposed to a specific magnetic field (negative or positive) absorbs and expresses the polarity of the magnetic field to which it is exposed while it is exposed. When the specific magnetic field exposure (negative or positive) is removed, the paramagnetic material has been charged and is a magnet possessing a north pole and south pole of equal gauss strength for a brief period of time.

When a weak magnet is exposed to a strong unipoled magnet, e.g., 2000 gauss magnet exposed to 12,000 gauss, the polarity of the weak magnet is overridden and the weak magnet assumes the polarity of the exposure from the stronger magnet during the exposure period. When the weak magnet is separated from the strong unipole magnet, it returns to being a single magnet with two poles of equal strength which have been magnetically charged to gauss strength capacity.

Cells are weak magnets and when exposed to the single pole of a strong magnet (300 gauss or more) assume the single pole of the stronger magnet. When the stronger external magnetic field is removed, the human cell resumes its polarity (two poles) which now have equal gauss strength and also have been magnetically strengthened to capacity. Thus it can be understood that human cells are magnetically strengthened by exposure to a single negative magnetic pole which exlains energy recover from sleep and relaxation.

Simultaneous exposure to both negative and positive magnetic fields of low gauss strength has been observed to evoke the same biological anti-stress response as exposure to the negative magnetic field alone. No one seems to know for sure the stress vs. anti-stress level of simultaneously applied positive and negative magnetic poles. The Germans believe the combined magnetic fields are anti-stress up to 200 gauss, the French up to 450 gauss and the Japanese up to 850 gauss. One thing we can depend on--negative magnetic energy is anti-stress no matter how long the exposure, whereas a brief exposure of both poles simultaneously can initially be anti-stress (negative magnetic energy effect) with prolonged exposure becoming stressful (positive magnetic energy effect).

The fact that chronic exposure to low (but beyond human adaptation) gauss strength of both magnetic poles can produce the positive magnetic pole effect

exlains how magnetic fields from exposure to low gauss strength electromagnetic currents can cause symptoms. Single negative magnetic field exposure counters environmental electromagnetic contaminates.

Negative magnetic field energy heals as well as normalizes human cellular metabolic function while at the same time disordering microorganism metabolism.

Negatively charged cellular DNA pulls oxygen (paramagnetic) into the cell since the counterclockwise rotation of negative magnetic field pulls magnetic or paramagnetic substances.

Alkaline media conserves oxygen and, therefore, the alkalinizing effect of negative magnetic field energy is a necessary part of tissue oxygenation.

Allergies, intolerances, chemical hypersensitivities, insect stings and most toxins are acids or biologically acid evoking and hypoxic producing. Hence, the alkalinizing and oxidizing effect of negative magnetic field energy is at least a substantial part of the answer for many human illnesses.

Effects of Negative Magnetic Energy on Living Metabolic Systems

Having addressed how magnetic energy works in general, how magnetic energy works in particular on the human body should be reviewed.

All injured and many diseased tissues are initially measurably magnetic positive but do not stay positive very long (3 hours for a broken bone) since the body concentrates negative magnetic field energy at the site for healing. Positive magnetic field energy interferes with healing and also drives abnormal cellular division and microorganism growth (infection). Negative magnetic energy is the energy that heals, alkalinizes, oxygenates and normalizes cellular metabolism in general. Therefore it is a major contribution to correcting disordered human metabolism. Negative magnetic energy:

Oxygenates Tissues

Cells are weak magnets with both a negative and positive pole. When a magnetic field is applied to a cell, the cell assumes the polarity of the applied magnetic field until the field is removed. Thus, when negative magnetic energy is applied to a cell, the counterclockwise spin of the negatively charged cellular DNA pulls oxygen (paramagnetic) into the cell.

Fights Infection
All microorganisms (bacterial, viral, fungal) and parasites are positive magnetic energy driven and are, therefore, stunned or retarded by negative magnetic energy. Two mechanisms are likely at work producing this effect: l) Negative magnetic energy alkalinizes (normalizes pH) tissues. Microorganisms and parasites flourish in acid media but are inhibited in alkaline media. 2) Negative polarity increases cellular oxygenation. Evidence indicates that negative magnetic energy charges up cellular DNA pulling oxygen into cells. (The counterclockwise rotation of negative magnetic energy pulls oxygen into the cells). Oxygen, needed for human cell metabolism, is destructive to micro-organisms and parasites which have a low tolerance for oxygen. Parasites are immobilized and therefore cannot feed while in a negative magnetic field.

Reduces Fluid Retention There are two types of edema 1) intercellular (inside cells) and 2) extracellular (outside cells). Intercellular edema clears rapidly under direct negative magnetic therapy which normalizes cell function of the sodium pump--sodium leaving the cell and potassium entering the cell. The fluid leaving the cell enters the extracellular space (blood and lymph system) and is carried away; thus the edema subsides. Extracellular inflammatory edema of closed spaces without drainage capacity (eye, sinuses, joints, etc.) also exists. In this instance negative magnetic energy placed to the side of the affected area will "pull" or drain the fluid from the encapsulated area relieving pressure and edema.

Encourages Deep Sleep While during waking hours the brain and central nervous system have a positive magnetic energy drive, during sleep the polarity of the brain and central nervous system swtiches to negative magnetic energy drive. Placing the body (head in particular) in a natural negative magnetic field setting encourages deep sleep (the switching of brain polarity). In addition, the pineal gland (a magnetic organ) is affected. This gland is responsible for production of melatonin which regulates the activity of the brain and the cyclic pattern of sleep and wakefulness. Melatonin is a master controller of energy expression during the waking hours.

Supports Biological Healing The healing process is affected by many of the conditions discussed here such as oxygenation of tissues, inflammation, in-fection, edema, etc. Negative magnetic energy is the "healing" energy of the Earth and discourages inflammation, infection and/or edema while encouraging healing by pulling oxygen and nutritional fluids into the injured area for speedy recovery.

Reduces Inflammation Inflammation, caused at different times by different conditions, is reduced by negative magnetic energy. Inflammation may be

caused by a bee sting (acid) in which case negative magnetic energy (alkali) neutralizes the acidity and relief is attained. Or inflammation may be associated with edema, infection, etc., which are also reduced by negative magnetic energy. (Refer to specific conditions described herein.)

<u>Normalizes Acid Base Balance</u> Negative magnetic energy is alkalinizing (never over-alkalinizes) and thus normalizes pH relieving symptoms resulting from acidic conditions: allergies, toxic states, acid indigestion, insect bites, sore/stiff musculature after athletic competition/exercise, chemical hypersensitivity and many disease states. An alkaline media conserves oxygen and therefore the alkalinizing effect of negative magnetic energy is a necessary part of tissue oxygenation.

<u>Relieves/Stops Pain & Other Symptoms</u> Pain, soreness and other symptoms are the result of infections, acidic states, inflammation, oxygen deficits, etc., all conditions improved by application of negative magnetic energy. Symptoms are relieved by treating the CAUSE. Headaches, a common complaint, caused by different things (acidic states, sinus pressure, allergic reactions, etc.) are likewise relieved by treating the CAUSE.

<u>Promotes Mental Acuity/Reasonableness & Bright Affect</u>
Negative magnetic energy l) influences the pineal gland responsible for secretion of melatonin which regulates every system including the brain, 2) normalizes abnormal electrial firing of the amygdala of the brain, 3) calms electrically overcharged brain function (hyperactivity, behavioral problems, obsessive-compulsiveness, schizophrenia, seizures), etc.

<u>Reduces/Dissolves Fatty Deposits</u>
Fatty tissue is acid. North pole magnetic energy being alkalinizing neutralizes and dissolves the fatty substance. This is a slow process.

<u>Aids in Reducing Fat & Calcium Deposits in Circulatory System</u>
Negative magnetic energy normalizes the acid base balance (pH) of the undesirable elements (cholesterol, triglycerides, etc.) which adhere and build up inside artery walls. In addition, with the application of negative magnetic energy, magnetic waves penetrate the tissues producing secondary currents. When the currents and magnetic waves clash, they produce impacting heat on the electrodes in the body cells. Hereby movement of hemoglobin in blood vessels is accelerated with calcium and cholesterol deposits in the blood decreasing along with other undesirable materials inclined to adhere to the inside of blood vessels.

SIDE EFFECTS

No adverse side effects have been observed with negative magnetic energy, the healing energy of the Earth. However, if pain or pressure does increase with application, the magnet should be moved in position but not necessarily removed. Pain can be caused by a pressure from fluid buildup within a joint or other encapsulated area (knee cap, sinuses, eyeball, etc.). Since negative magnetic energy pulls additional fluid into the enclosed area, pain can increase if direct application is made to an area already full of fluid with no drainage capacity. In this instance, moving the magnet above, below or aside the painful site 2-3" will pull fluid away from the affected area thus relieving pain.

Persons on medications (prescription drugs, thyroid, etc.) may develop symptoms of overmedication when using the overall body treatment (magnetic bed). As stress and inflammation are reduced in the body, toxins are released and metabolism improves. Drugs can become more effective and may have to be reduced or discontinued as body chemistry changes. Consult a physician for monitoring if symptoms occur.

PRECAUTIONS

DO NOT use magnets on the abdomen during PREGNANCY.

DO NOT use magnetic bed or magnet(s) on chest with PACEMAKER or defibillator. A magnet within two inches of these instruments can interrupt function.

DO NOT not use magnetic bed 24 hours daily if ill. Adrenal function may be suppressed with slow energy recovery. Magnetic bed may be used 8-10 hours per day and then supplemented with magnetic spot therapy for symptoms such as for pain, inflammation, etc.

DO NOT use magnetic energy on abdomen for 60-90 minutes following meals in order to allow peristalsis to take place.

DO NOT use positive magnetic energy unless under medical supervision. Positive magnetic energy can l) overstimulate brain activity producing seizures, hallucinations, insomnia, hyperactivity, addiction, etc. and 2) stimulate growth of tumors, microorganisms, etc.

ROUTINE PREVENTIVE MAGNETIC THERAPY

Pondering the lifestyle of the l990s with fast foods, nutritionally deficient diets, sedentary jobs, electromagnetic exposures, high stress occupational and academic loads, multiple family pressures, etc., effective and easily applied preventive measures are certainly called for to keep physical and mental function at optimum. Negative magnetic energy fits all the requisites and goes even further.

Magnetic Bed

For preventive measures and/or treatment of symptoms and diseases resulting from above mentioned stressors, total body treatment with a magnetic bed provides the best overall benefit. The magnetic bed places the entire body in a negative magnetic field all night--a natural sleep inducing, stress reducing, anti-inflammatory environment. The health rewards of restorative sleep and stress management cannot be overemphasized.

Although total body treatment is preferrable, if a magnetic bed is not available, the following Nightly Routine may be observed:

Routine Nightly Treatment

<u>Thirty to sixty minutes before retiring for the night, apply negative magnetic energy as follows:</u>

a) HEAD: 3X6" plastiform magnet + round neodymium magnet placed on low occipital area (back of head/base of skull where skull joints soft neck tissue) Alternative: 4X6" ceramic magnet under head while reclining.
b) SPLEEN: 4X6" flat ceramic magnet (L side low back at junction of rib-cage/abdomen
c) LIVER: 4X6" flat ceramic magnet (R side low back junction of rib-cage/abdomen)
d) THYMUS: 2X5" ceramic or 3X6" plastiform magnet (anterior mid chest)
e) NECK LYMPH NODES: 2X3" plastiform magnets over lymph nodes on upper anterior aspect of neck under chin. Alternative: 2 cube magnets one on each side of neck or l cube magnet on one side + (2) 2X2" plastiform magnets stacked on opposite side.

<u>Upon retiring for the night:</u> Negative magnetic energy may be applied at the crown of the head all night during sleep using 4 4X6" ceramic magnets standing upright l" apart--the crown of the head (top) resting 2-3" from the magnets during sleep placing the head in the negative magnetic field all night. <u>Alternative:</u> Apply negative magnetic energy by a round Neodymium to the crown of the head (exact center top of head) all night. This has the potential to

enhance deep sleep, increase physical and mental energy and treats the pineal gland responsible for production of melatonin which controls the cyclic pattern of sleep and wakefulness.

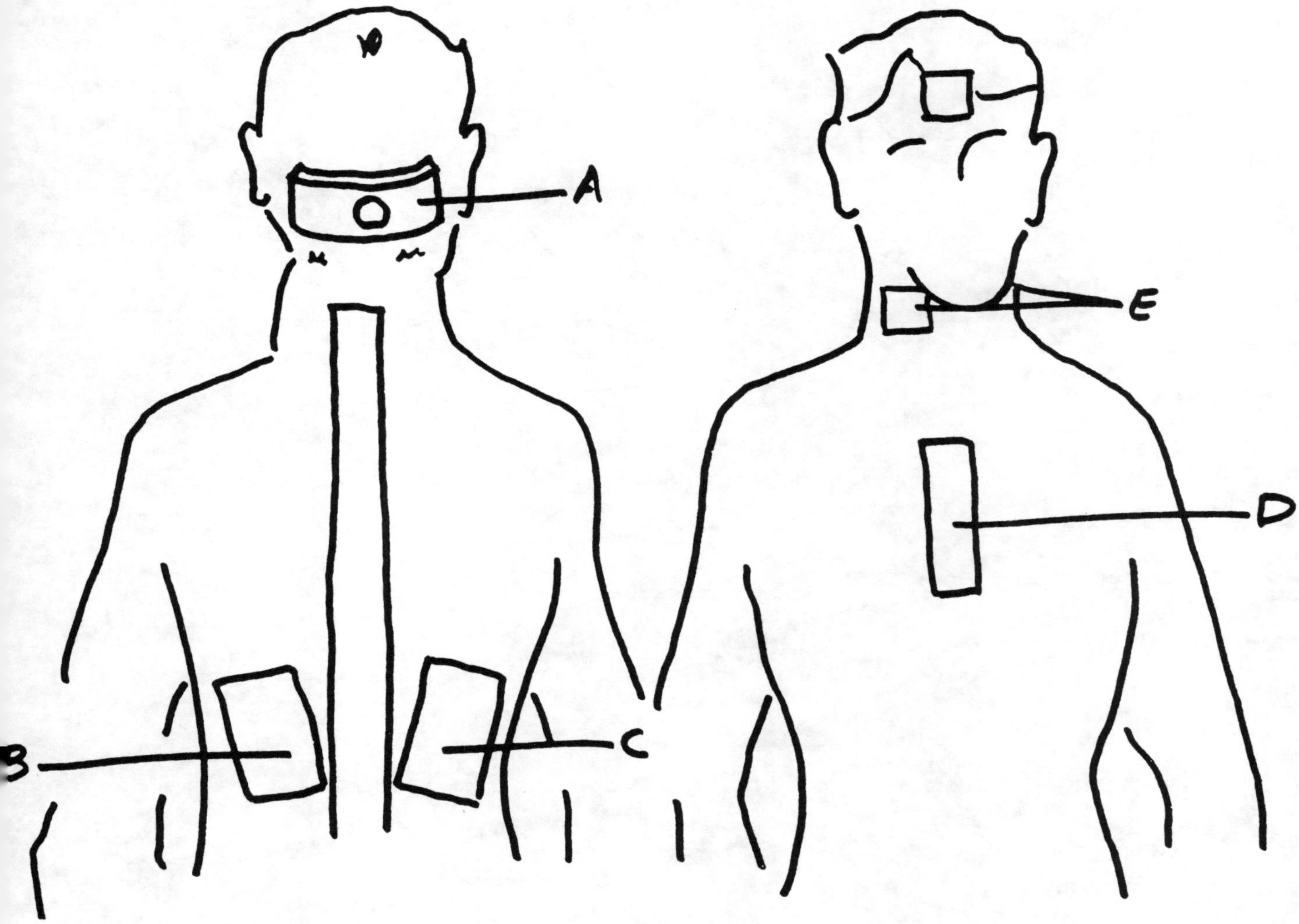

Educational only. Not to be construed as prescribed medical treatment for any individual. If you have a health problem, consult your physician.

Magnetically Treated Water

The treatment and drinking of magnetically treated water may be of additional preventive and therapeutic value. With the drinking of magnetically (negative) treated water, the beneficial effects of negative magnetic energy are passed on to the biological system. Water treated with negative magnetic energy is calming.

To treat water:

l) Place water in glass container
2) Set container on negative pole of a magnet
   OR
   Tape magnets to container with negative pole directed towards contents
3) Treat for minimum of 5 minutes, preferrably longer

Drink a minimum of 2 8-ounce glasses daily. During flu epidemic, drink magnetically treated water every 4 hours for prevention or relief of symptoms.

For quicker energy pick-up in the morning, drink l 8-ounce glass of bipolar magnetically treated water. To treat water:

l) Place water in glass container
2) Place negative pole of a magnet on one side of filled jar and positive magnetic pole on opposite side
3) Treat water for a minimum of 5 minutes before drinking

DISORDERS

The following research information on magnetic therapy for various disorders is compiled for educational purposes only and is not to be construed as prescribed medical advice in any specific situation. If symptoms are present, consult a physician. Magnetic therapy refers to NEGATIVE magnetic energy only!

Variables in magnetic therapy include:

1) gauss strength
2) negative vs. positive magnetic energy
3) placement and
4) duration of exposure

The protocols herein were found to be the most effective in a wide range of research subjects but placement, gauss strength, and duration of exposure needed for individual relief may vary from person to person.

ACNE

Any size magnets described herein may be applied directly to affected areas. In addition, the Nightly Routine may be beneficial. Often acne is caused by food allergies, especially to dairy products, which should be tested per testing instructions in Brain Allergies: the PsychoNutrient Connection
by W.H. Philpott, M.D.

ADDICTIVE STATES (Drugs, Chemicals & Foods)

Addiction is a state in which a drug, food or chemical containing a narcotic (exorphin) produces a pleasurable brain reaction, relieves physical symptoms and in general soothes both physically and emotionally.. In addition, frequently used foods and chemicals can evoke the body to produce self-made narcotics (endorphins). When endorphin production exceeds the body's physiological need for endorphins, they become as addicting as exorphins. Thus food addiction is a very real narcotic addiction and follows all the rules of biological responses to narcotics including, of course, a miserable set of withdrawal symptoms if the food is not eaten regularly. The most common addictive substances are caffeine, wheat, corn, sugar, beef, nicotine plus
favorite foods or drugs which are used often.

Eating disorders (compulsive urge to overeat, anorexic vomiting, purging and anorexic loss of appetite) have all been observed to relate to food addiction as well as IgG food allergy. The antibody release phenomenon of IgG food allergy occurring after the food is digested provides an addictive withdrawal phase the same as addiction with the initial rise and later drop in endorphins &/or exorphins.

Addiction produces positive magnetic field stimulation and overdrive of the brain and body. Energy restoration achieved by negative magnetic energy maintains health while positive magnetic energy overdrive (stimulation) eventually leads to burn-out of metabolic functions and a collapse of health. The answer to reducing the positive magnetic field overdrive is to provide sufficient negative magnetic energy to normalize the magnetic balance of the body and brain. Negative magnetic energy will not only aid in stopping the addictive craving for the substance but may relieve addictive withdrawal symptoms.

Withdrawal symptoms and eating disorders have responded with the following:
- Sleep on magnetic bed nightly
- Repeat Nightly Routine and/or CVS therapy as often as necessary to achieve comfort level (day or night)
- Apply negative magnetic field directly to areas of discomfort: head, stomach, etc., using any magnet(s) described herein
- 2X1 3/8" cube magnet placed on right & left temple just above & in front of the ears may be needed to calm emotional state during withdrawal phase
- Sleep with negative magnetic energy at crown of head nightly as described in Sleep

ALCOHOLISM

Magnets, preferrably ceramic magnets, may be applied to liver, pancreas, diaphragm, CVS. In addition, the Nightly Routine may add benefit. Food allergies should be examined as described in Brain Allergies: the PsychoNutrient Connection by this medical author.

ALLERGIC REACTIONS

Immunologic reactions may be general involving spleen, thymus, lymph nodes, etc. or may be tissue specific affecting sinuses, throat, lungs, GI tract, skin, eyes, brain, etc. In addition to the usual helpful applications of avoidance, immunization, injection therapy and nutrition, magnetic energy may help normalize the immune response, reduce swelling and offset allergic inflammatory response.

General research treatment:
- Magnetic Bed or Nightly Routine
- Twice daily for 30 minutes:
  4X6" ceramic magnet to spleen
  2X5" or 4X6" ceramic magnet to thymus (immune system)

Specific research treatment (magnetic energy applied specifically to tissues or organs involved):

• Nasal & sinus: Negative magnetic application for 30 minute periods 2 or more times daily with any size plastiform magnet or 2X1 3/8" ceramic cube magnet(s). If drainage of involved sinus is blocked, place magnet above, below or to the side of the sinus to pull fluid from the sinus. Pain could increase in a blocked area of the sinus if magnet pulls more fluid into closed area. At times placing the magnet as far away as the neck may favorably influence the sinuses in the face.

• Skin reaction: Direct application over affected area with any size plastiform, ceramic or neodymium magnet

• Gastrointestinal reaction: Food Reaction instructions
The motility of the stomach and intestine is stopped during the period of exposure to the magnet. Therefore, it is preferrable to wait 1-2 hours after eating to treat the abdomen. This lapse of time allows peristalsis and digestion to occur. However, if gastrointestinal symptoms occur post meal, treat at any time for a brief period sufficient to relieve symptoms.

• Eye reactions: Negative magnetic application 15-30 minutes 2 or more times daily by placing any size plastiform or 2X5" ceramic magnet directly over eyes--with the EXCEPTION of glaucoma patients. Refer to Glaucoma.

<u>ANKLES</u> (swollen)
Magnetic energy application to kidneys and lymphs using any size magnets described herein. In addition, Nightly Routine may be followed.

<u>ANXIETY</u> See l) Depression and 2) Emotional and Mental Disorders.

<u>APPENDICITIS</u>
Contact physician.

<u>ARTERIOCLEROSIS</u>
See Atherosclerosis

<u>ARTHRITIS</u>
The Magnetic Bed, with the potential to reduce inflammation, swelling and pain, provides optimum magnetic therapy for arthritic disorders including gout and ankylosing spondylitis. In addition, specific areas of discomfort have responded to the following application.

Since pain can be caused by different conditions, the CAUSE of the pain must be determined by self-test:
l) Pain can be caused by a pressure from fluid buildup within a joint or

encapsulated area. Since negative magnetic energy pulls additional fluid into the treated area, pain can increase if direct application is made to a joint already full of fluid. In this instance, moving the magnet above, below or aside the painful site 2-3" may pull fluid from the affected area relieving pain.

2) Pain can also be caused by inflammation. In this instance direct application of magnetic therapy to the painful area may provide relief by reducing swelling and inflammation.

If a Magnetic Bed is not available, the Nightly Routine may be followed regularly along with the above instructions for direct treatment of specific areas of inflammation.

<u>ASTHMA</u>

Although magnetic therapy has been demonstrated to provide relief in a number of instances, it is recommended with acute bronchospasm that a physician be contacted.

A female with a history of bronchial asthma when suffering an acute bronchospasm received relief with the application of a 4X6" ceramic magnet over the sternum. Relief began after 10 minutes exposure and was complete within 30 minutes. Negative magnetic energy is mildly constricting and for this reason may increase an acute bronchospasm. In such case the magnets are applied up to the tolerated limit, removed and reapplied frequently in order to relieve inflammation and edema.

<u>ATHEROSCLEROSIS</u> (Hardening of the Arteries)

A thickening of the inner lining of the arteries (blood vessels that carry oxygen and other nutrients from the heart to other body parts). Atherosclerosis may lead to kidney damage, decreased circulation to brain and extremities and coronary-artery disease.

Circulating cellular elements in the blood and blood cell walls are normally negative magnetic poled. All blood elements being of the same polarity (negative) prevents sticking and clumping of the separate elements. When the pH of the blood drops lower than normal (7.4 pH) then fatty acids (positive) begin to stick to the walls of arteries (negative). As fatty deposits accumulate, they reduce the blood vessel's elasticity and narrow the passageway interfering with blood flow.

Negative magnetic energy is calculated to reinforce or restore the natural negative pole magnetic state of blood cells and blood cell walls. Not only does

the natural negative polarity of the blood prevent the formation of plaque but negative polarity is necessary to prevent inflammation in the blood and blood vessel walls. Since atherosclerosis develops when the negative magnetic field is reduced, application of the negative magnetic field over an artery restores normal magnetic field balance and thus helps prevent and/or reverse the disease process.

In addition, negative magnetic energy has been observed to aid in the removal of calcium deposits from arterial atherosclerotic plaques as well as rheumatoid joints.

Maximum treatment: Total body treatment with a Magnetic Bed 8 hours nightly while sleeping treats the entire blood circulatory system of the body.

Specific treatment: (when vascular studies are available) Specific areas of major atherosclerotic plaques may be treated directly with 4X6" ceramic magnets. In addition, a 3X3" or 2X2" plastiform magnet may be worn over the heart (left chest area) in shirt pocket or bra cup during waking hours.

General treatment (without benefit of vascular studies)
Minimum treatment: 30 minutes daily as follows
Ideal treatment: 30 minutes 2-4 times daily as follows

- 4X6" ceramic magnet at Circle of Willis at base of skull (placed under a pillow)
- 4X6" ceramic magnet on bifurcation of the aorta--right and left side in groin area where legs join body (pubic hairline)
- Carotid arteries and base of brain
- Specific symptomatic areas may be treated directly, e.g., dry eye caused by arteriosclerosis of brain
- Arteries of the legs may be treated by using 3X24" plastiform magnets the entire length of both legs on posterior & anterior (optional)

In addition, during waking hours 2X2" plastiform magnet may be worn over heart (left chest area) in shirt pocket or bra cup during waking hours.

Excesss fat is eliminated from the body through the liver-bile-gallbladder system. Since this elimination process requires taurine, it is advisable to supplement taurine 500 mg 2-3 times daily. In addition it may be beneficial to support the liver-gallbladder processing of fats by exposing the liver to negative magnetic energy by reclining on a 4X6" ceramic magnet directly under the liver or placing a magnet directly over the liver. Of course, total body treatment does provide liver exposure to negative polarity.

In addition to magnetic therapy, a medical evaluation should be obtained to

evaluate the cause of the fatty acid buildup in the system perpetuating the atherosclerotic process in the first place. Additional beneficial therapies are available such as oxidative therapy, nutritional supplementation and EDTA chelation.

BACKACHE
Lower backache has been most effectively relieved by placing magnetic energy down the spine with a 24" plastiform magnetic strip and also treating the kidneys with 2 4X6" ceramic magnets. If discomfort persists, bladder/kidney infection should be considered. Infection requires vigorous magnetic treatment and may also require antibiotics. A physician should be consulted.

BEDWETTING
Bedwetting is often caused by food allergies. However, the condition may be helped by using the Nightly Routine plus negative magnetic energy applied directly to kidneys and ureters with either ceramic magnets or 3-4 stacked plastiform magnets described herein.

BEHAVIORAL DISORDERS See Emotional & Mental Disorders.

BITES, INSECT
Insect bites and stings should be treated as quickly as possible using any size magnet described herein. Bites and stings are acid. Negative magnetic energy reduces acidity and inflammation thereby relieving painful symptoms.

BLADDER PROBLEMS
Bladder and kidney problems have responded by direct application of negative magnetic energy with ceramic or 3-4 stacked plastiform magnet(s) described herein. In addition the Nightly Routine was followed.

BRAIN ALLERGIES
"Brain allergies" are caused by allergies or hypersensitivities to foods, chemicals, etc, which produce a swelling in the brain evoking symptoms such as mental fatigue with poor concentration, headache, depression, lethargy or hyperactivity, etc.--and even psychosis in severe cases. Different head placements of magnetic therapy need to be tried for best results: CVS, forehead, bitemporal areas plus the spine. Refer to Brain Allergies: the PsychoNutrient Connection, W.H. Philpott, M.D.

BREAST (lumps)
Negative magnetic therapy should be applied directly to the affected areas using ceramic magnet(s) or 3-4 stacked plastiform magnets described herein. In addition, the Lymph System should be treated. DO NOT USE MAGNETS ON CHEST WITH A PACEMAKER OR DEFIBRILLATOR!

BRONCHITIS
The chest/lung area may be treated directly with negative magnetic therapy using ceramic or 3-4 stacked plastiform magnet(s) of any size described herein.

DO NOT USE MAGNETS ON CHEST WITH A PACEMAKER OR DEFIBRILLATOR!

BURNS

Apply negative magnetic energy as soon as possible to burns before tissue deterioration occurs. Apply any size magnet readily available--preferrably a 2X5" or 4X6" ceramic magnet.

A man burned his finger with steam which was extremely painful. For 10 minutes the finger was packed in ice....at 15 minutes a blister was forming. At this point the finger was placed on the negative side of a 4X6" magnet. Within 15 minutes the blister had resolved. The pain gradually subsided and within one hour of magnetic exposure there was neither pain or blistering. Two weeks later the superficial skin peeled leaving normal new skin.

A woman scorched the ends of her fingers testing the burner on an electric stove and the fingertips quickly turned white and painful. A magnet was fortunately close by and she immediately placed fingertips on the negative pole. Within 5 mintues the pain was gone and within 15 minutes a normal pink color returned to the scorched area. Blistering and peeling did not occur.

BURSITIS

Often bursitis is caused or aggravated by food allergies. Dietary habits should be examined. Painful areas of bursitis may be treated directly with magnets of any size described herein, preferrably ceramic magnets.

CALCIUM DEPOSITS
A Magnetic Bed providing total body therapy may be beneficial. In addition, areas of calcium deposit should be treated directly with magnetic therapy using any size magnet(s) described herein. Special attention should be given to the parathyroid (thyroid area) also.

CANDIDA ALBICANS Refer to Fungal Infection.

CAROTID ARTERIES
Carotid arteries are treated by placing 2 ceramic cube magnets on each side of the neck anteriorly. 2X5" ceramic magnets or 3-4 stacked plastiform magnetic strips may also be used.

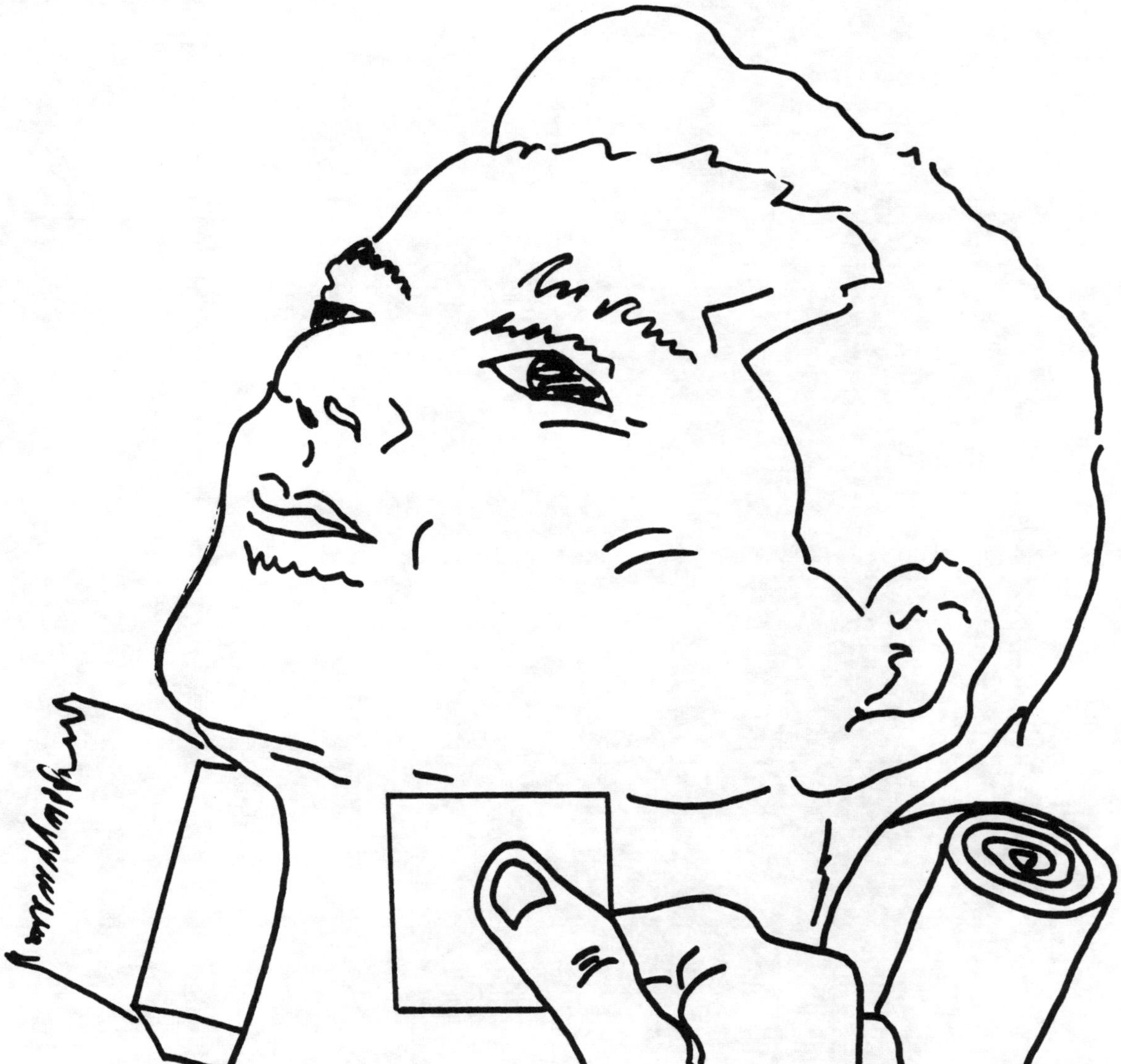

Educational only. Not to be construed as prescribed medical treatment for any individual. If you have a health problem, consult your physician.

CATARACTS

Cataracts are the result of a lack of sufficient oxygen to the eyes. Negative magnetic energy applied daily to the eyes has been demonstrated to reverse some cataract formation as well as aid in preventing further development of cataracts.

Eyes may be treated directly by placing the negative pole of a plastiform magnetic strip (any length) over the eyes for 30-60 minutes three times daily. Preventive measures would indicate applying magnetic energy as described for 30 minutes daily.

CELLULITE DEPOSITS

Cellulite should be treated by placing ceramic or stacked plastiform magnets directly on fatty tissues. Magnets should be bound to affected areas such as thighs, buttocks, upper arms, etc., with an Ace Bandage or other body wrapper and held in place nightly during sleeping hours. Negative magnetic energy dissolves the fatty tissue under treatment. It is to be noted that fat dissolution is slow and it will take a period for results to be obvious.

Abdominal fatty deposits may be treated using 3 stacked 4X12" plastiform magnets held on the fatty area with a binder during sleeping hours. This is compatible with the magnetic bed. Fluid and fecal matter are pulled under the magnet and, therefore, when treating the upper abdomen, moving the magnet to the lower abdomen for the last two hours of the night will help prevent constipation. Treating the lower abdomen encourages a bowel movement upon arising in the AM. This application should also be considered for intestinal parasites and candida infection.

CEREBELLAR VESTIBULAR SYSTEM (CVS)

The CVS acts as a computer center sorting l) function and coordination of sensory inputs from the body to the brain (cerebral cortex) as well as 2) motor function and other messages from the brain to the body. The sense of "fright and flight" in the CVS are responsible for the development of phobias and compulsions. The Cerebellar Vestibular System incorporates the brain stem, inner ear connections and cerebellum. The cerebral cortex is the computer center controlling memory, imagery, fantasy, reasoning, integration of memorized material, and initiation of motor function. Certain areas also govern sensory and visual function.

The CVS (low occipital area upper neck) is treated by using ceramic magnet(s), 3-4 stacked plastiform magnetic strips or a plastiform magnetic strip with a round Neodymium magnet stacked (illustrated). To some degree the magnetic energy

radiates into the base of the brain producing a calming effect. Since this is the integration area between body and brain, for best results the CVS area should be treated simultaneously when treating other body areas.

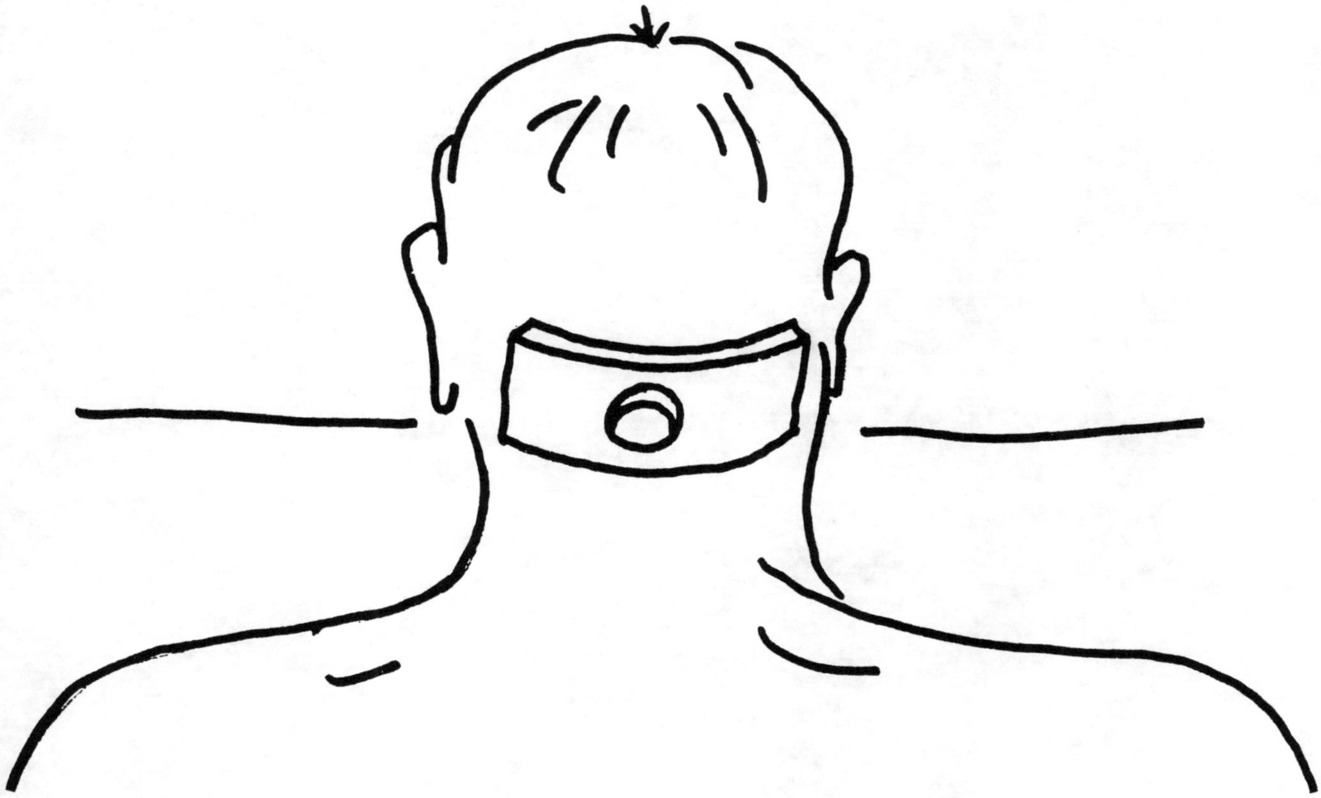

Educational info only. Not to be construed as prescribed medical treatment for any individual. If you have a health problem, consult your physician.

CHEMICAL EXPOSURES/REACTIONS

Reactions occurring from exposure to chemicals while shopping, traveling, working, etc., have been treated successfully with the Nightly Routine, the CVS placement and particularly the liver placement. Also refer to treatment for Brain Allergies.

The Magnetic Bed offers value in this instance. If Magnetic Bed is not available, use the Nightly Routine NIGHTLY for chronic cases with residual effects from chemical exposure PLUS other placements described in this book for relief of specific symptoms. See Chemical Exposures in introductory chapter.

CHILDHOOD DISEASES

Childhood diseases may be treated as described in Viral Infections. In addition, direct application of magnet(s) described herein may be made to painful and irritated areas.

CHOLESTEROL & TRIGLYCERIDES

Cholesterol and triglycerides are composed of fatty acids. Negative magnetic energy alkalinizes (normalizes pH) and thus neutralizes fatty acids. Therefore, cholesterol and triglycerides can be reduced by the application of negative magnetic energy. Theoretically, the rise in blood fats occurs because of an imbalance between negative and positive magnetic energy in which there is an ascendancy of positive magnetic energy. Proper balance between negative and positive energy could prevent the development of the problem.

Total body treatment is ideal because blood circulating throughout the entire body is treated. The easiest method in treating the entire body is by sleeping on a Magnetic Bed 8 hours nightly.

If a magnetic bed is not available for total body treatment, it would be advisable to follow the Nightly Routine and in addition selectively treat the thyroid (controls cholesterol levels) and liver (manufactures cholesterol) with direct application of either ceramic or 3-4 stacked plastiform magnets.

In addition, negative magnetic energy may be applied over the heart for as many hours as desired treating the blood as it passes through the heart (See Heart). Negative magnetic exposure to the heart has not been observed to slow the heart rate more than 10 beats/minute with adapation and adjustment soon made in the heartbeat. This has not been observed to pose any problem.

NOTE: Magnetic Bed or magnet(s) on the chest should not be used with a Pacemaker or defibrillator!

David, 46, had a history of extremely high cholesterol and triglycerides plus a family history of blood fat problems. On a routine physical exam blood fats were alarming abnormal and he was considered at extreme high risk. Results as follows:

Initial physical		
Triglycerides	1490	Normal: 19-175
Cholesterol	Elevated	Normal: <200
HDL	Too turbid for a result!	Normal: 34-65

The patient was placed on Lopid and a special diet and remained on this treatment for two years with lowering of blood fats as follows:

Triglycerides	725
Cholesterol	Normal
HDL	24

At this time the patient was placed on a Magnetic Bed. After sleeping on the Magnetic Bed for six weeks and reducing Lopid to half dosage, results showed:

Triglycerides	213
Cholesterol	Normal
HDL	28

Albert, age 80, was placed on a full magnetic bed consisting of 2000 gauss strips only. Laboratory results were as follows:

	Before Magnetic Therapy	After 30 days on Magnetic Bed
Triglycerides	170 (19-175)	123
Cholesterol	263 (<200)	262
LDL	190	182
HDL	42 (34-54)	47

Helen, age 72, experiencing elevated blood fats was placed on 850 gauss magnetic bed with vibratory frequencies. Monitoring after a thirty day trial period revealed:

	Before Magnetic Therapy	After 30 days on Magnetic Bed
Triglycerides	255 (19-175)	174
Cholesterol	220 (<200)	207
LDL	136	134
HDL	35 (34-54)	38

<u>CHRONIC FATIGUE SYNDROME (CFS)</u>

Very frequently chronic fatigue is caused by the presence of viral infections. The most effective treatment method would be total body treatment with a Magnetic Bed. If Magnetic Bed is not available, magnetic therapy should be applied as described herein for Viral Infections. The Nightly Routine should also

be beneficial.

CIRRHOSIS OF LIVER

Magnetic therapy should be applied directly to liver, pancreas and glands using ceramic magnets or 3-4 stacked plastiform magnets described herein.

COLDS

Treat colds the same as Viral Infections along with local treatment of chest/lung, throat, CVS, lymphs. Preferrably use ceramic magnets or stacked plastiform magnets as continuously as possible until symptoms begin to subside. Treatment then may be reduced.

DO NOT USE MAGNETS ON CHEST WITH PACEMAKER OR DEFIBRILLATOR!

COLITIS

Negative magnetic therapy should be applied directly to the abdominal area including specifically the liver, gallbladder and colon using 4X6" ceramic magnets or 2-3 stacked 4X12" plastiform magnets. Treat as long as necessary or until symptoms subside. Refer to Gastrointestinal Disorders.

A 65 year old man with chronic colitis produced by cytomegalovirus was starving and dehyrated from frequent diarrhea. The illness was so advanced that he required intravenous tube feeding. A plastiform magnetic strip was placed on the abdomen which stopped peristalsis and thus the diarrhea. In fact, the magnet had to be removed in order for the patient to have a regular bowel movement. Gradually the magnet was removed for brief periods of time until finally the patient could go several hours without diarrhea. The patient's condition remained well controlled with magnetic energy application one hour three times a day.

CONCENTRATION, POOR

"Mental cobwebs" have been successfully cleared from the brain by applying a plastiform or neodymium magnet to the forehead for 15-30 minutes. In some instances it may be necessary to use the CVS placement or place 4X6" ceramic magnets 1-2" from the crown of the head for a brief period of time.

CONGESTION, LUNG

For congestion affected lung(s) may be treated directly with 4X6" ceramic magnet(s). In addition, use the Nightly Routine.

DO NOT USE MAGNETS ON CHEST WITH PACEMAKER OR DEFIBRILLATOR!

CONJUNCTIVITIS
The eye may be treated directly by placing any size magnet(s) referred to herein directly over the affected eye.

CONSTIPATION
For constipation magnetic therapy should be applied to the liver, gallbladder and colon areas. Sleeping at night with 3 4X12" plastiform magnets stacked on the abdomen retains fluid in the stool and is, therefore, anticonstipating. Hold magnets in place with an abdominal binder. Negative magnetic energy pulls fluid into the colon softening the stool and allowing natural elimination. Supplementation of Vitamin C (12 Gm daily) will also aid elimination.

COUGH
Place ceramic magnet(s) or 3-4 stacked plastiform magnetic strips over the chest, neck &/or throat area for relief.

CRAMPS, MENSTRUAL
For abdominal discomfort use 4X6" ceramic magnets over reproductive organs--uterus, ovaries and glands. Simultaneous treatment of the CVS can prove beneficial following illustration herein. The Nightly Routine practiced faithfully premenstrually may prevent development of severe symptoms. (For other specific symptoms, e.g. headache, refer to the appropriate section of this handbook.)

CROUP
Consult physician.

CYSTITIS
Use ceramic magnets on kidney, bladder and ureters 1-3 hours daily or continuously until symptoms subside.

DEPRESSION

Read Brain Allergies, the PsychoNutrient Connection, by W.H. Philpott, M.D. The Magnetic Bed may be useful in cases of depression.

Depression, anxiety and panic are all likely to be helped by raising melatonin (See Pineal Gland and Sleep, Method II) by sleeping with negative magnetic energy at the top of the head. In addition, trial placements of magnets should be used to achieve maximum relief such as Bitemporal Placement, CVS, and Nightly Routine. Also refer to Emotional Disorders.

Depressives also often benefit by supplementation of L-tyrosine 500 mg 3 times a day and niacinamide 500-1500 mg at bedtime or niacin 100-500 mg three times daily.

Some depressives require l) therapeutic bright light applied to the eyes for a few minutes at the beginning of each day and/or 2) several minutes of vigorous motor exercise each AM for arousal and energy initiating effect. In addition, some depressives also require mid-day 30-60 minutes of sleep in a dark room with negative magnetic energy at the top of the head followed by application of bright light, an exercise period and 500 mg L-taurine and 500 mg L-taurine again at bedtime.

For further information on the use of bright light:

- Czeisler, CA, et al. "Exposure to Bright Light and Darkness to Treat Physiologic Maladaptation to Night Work", New England Journal of Medicine, 322:18, May 3, l990
- Van Cauter, E and Turek, FW. "Strategies for Resetting the Human Circadian Clock", New England Journal of Medicine, 322:18, May 3, l990
- Parry, BL, et al. "Treatment of a Patient with Seasonal Premenstrual Syndrome", American Journal of Psychiatry, 244(6), June l987
- Jacobsen, FM, et al. "Seasonal Affective Disorder: a Review of the Syndrome and Its Public Health Implications", American Journal of Psychiatry, 77(l), January l987
- Wirz-Justice, A, et al. "Light Treatment of Seasonal Affective Disorder in Switzerland", Acta. Psychiatry Scand, 74(2), August l986
- Hellekson, DJ, et al. "Phototherapy for Seasonal Affective Disorder in Alaska", American Journal of Psychiatry, 143(8), August l986
- Rosenthal, NE, et al. "Antidepressant Effects of Light in Seasonal Affective Disorder", Archive General Psychiatry, 41(l), January l984

DIABETES

For management or possible reversal of diabetic process through food testing and proper dietary practices, refer to Victory Over Diabetes: A BioEcologic

Triumph by W.H. Philpott, M.D. Food testing should be medically supervised.

The Magnetic Bed provides overall body therapy needed in diabetes. If Bed is not available, follow Nightly Routine plus use magnetic therapy as described in this Handbook for specific complications of diabetes, e.g., pain, food reactions, etc. Treatment should be applied a minimum of 3 hours daily for any specific problem.

DIARRHEA

Place ceramic magnet(s) or stacked 4X12" plastiform magnets on colon and mid lumbar area.

DIVERTICULITIS

For pain and discomfort place ceramic magnet(s) or stacked plastiform magnets directly on painful areas of colon PLUS on liver and gallbladder. May be used as often as needed and as long as necessary for symptom relief.

DIZZINESS

Since dizziness can be caused by different conditions, different magnetic placements must be tried until symptom relief is achieved. Blood pressure should be checked and, if high, instructions for Hypertension applied. The CVS placement on the head should be given a trial. Since indigestion can cause dizziness, 4X6" ceramic magnets should be tried on the stomach. For ear involvement, any size ceramic magnets or stacked plastiform magnets can be placed directly over the ears.

DRUG ADDICTION

See Addictive State (Drugs, Chemicals, Food).

DRY EYE

Dry eye is caused by arteriosclerosis of the brain. The eye should be treated directly with 2X5" ceramic, 3X6" plastiform or a neodymium magnet 30 minutes daily or several times daily as needed. Also treament for arteriosclerosis should be administered as described in this handbook.

EARACHE

Apply either ceramic magnets or plastiform magnetic strips to ears, throat, neck (Eustachian tube), 7th cervical and forehead to find areas which offer best relief. Magnets may be left in place until relief is obtained.

EDEMA

There are two types of edema l) intracellular, inside the cells and 2) extracellular--outside the cells.

Intracellular edema clears rapidly under direct negative magnetic therapy which normalizes cell function of the sodium pump. Sodium leaves the cell and potassium enters the cell. The fluid leaving the cell enters the extracellular space (blood, lymph system) and is carried away; thus the edema goes down. No matter what caused the cell edema (allergic reaction, toxins, burn, insect bite, bruise, etc.), the swelling is rapidly resolved. For chronic fluid retention, the Magnetic Bed with total body treatment offers optimum therapy. If Magnetic Bed is not available, treat lymph system, kidneys and the affected area with ceramic magnets or plastiform magnetic strips until relief is obtained.

Extracellular inflammatory edema of closed spaces without drainage capacity (eye, sinuses, joint capsules, etc.) should be treated with ceramic magnets or stacked plastiform strips 2-3" above, below or to the side of the affected site. Fluids will be thus drained from the inflammed site by the pull of the negative magnetic field.

Swelling of extremities (arm, leg) caused by reduced return flow of blood or lymph is a special problem which cannot be expected to respond to direct application of magnetic therapy. CONSULT A PHYSICIAN. The affected limb should be elevated to encourage drainage of blood/lymph plus ceramic magnets or stacked plastiform magnetic strips may be placed above blockage, e.g., front and back of shoulder in an affected arm or above groin in an affected leg, etc. If intracellular edema is also present, place magnets directly over the affected area and keep limb elevated for drainage. In case of blocked drainage from the limb, consideration should be given as to the cause of the blockage. If there is inflammation, scar tissue or injury at the site of the blockage, place ceramic magnets or stacked plastiform magnets directly over the blockage for resolution. A limb with limited blood supply due to atherosclerosis or blockage can develop inflammed or gangrenous areas which will not heal. Such areas should be treated directly with ceramic magnets or stacked plastiform magnetic strips. Again, CONSULT A PHYSICIAN.

ELECTROMAGNETIC INTERFERENCE

Available correction for the effects of electromagnetic contamination and

positive pole magnetic field exposure includes I) avoidance of exposure, if possible, and 2) application of sufficient negative magnetic pole energy to offset the disordering influence.

Negative magnetic energy correction for electromagnetic health stressors would include:

- Ideally a Magnetic Bed providing a protective shield of negative magnetic energy and natural earth frequency about the body during sleeping hours

OR

- Nightly Routine followed regularly if Magnetic Bed is not available
- During daytime hours negative pole of 2X2" plastiform magnet worn over heart (See Heart)
- Premeal negative magnetic exposure to the liver, spleen, sternum and back of head or forhead for 15-30 minutes with ceramic magnet(s) or 1-4 stacked plastiform magnetic strips (See "Food Reactions)
- Negative pole of magnet over sternum as a shield when exposed to computer screens, TV, etc.

A man was in close proximity to a lightning strike which knocked him several feet through the air. After the shock he was extremely weak. Ceramic magnets were placed at the crown of his head and under him during one night's sleep. By morning his energy had returned and he was symptom-free upon awakening. This is an example of the ability of negative magnetic energy to correct the adverse effects of electromagnetic contamination.
(Also refer to Electromagnetic Contamination).

EMOTIONAL & MENTAL SYMPTOMS & DISORDERS

Optimum physical and emotional health consists of a balance between excitation and inhibition. In no place is this more obvious than in the central nervous system (CNS) where there are periods of wakefulness, alertness and activity. However, these states of excitation cannot be maintained for extended periods of time without dire metabolic consequences wherein organs become overstimulated and wear out. Periods of rest, relaxation and sleep are absolutely necessary in order for metabolic processes to rejuvenate and restore the energy of the body. Thus, physical and emotional health are a BALANCE of excitation (positive magnetic energy) and inhibition (negative magnetic energy).

Magnetic energy consists of two poles--negative (inhibition) and positive (excitation). As has been previously discussed, magnetic energy influences the metabolic system in specific directions. Thus the two opposite magnetic energy

drives can be used to keep biological systems in a functional balance of excitation and inhibition.

Much human misery results from extending the excitation of the biological system beyond its normal capacity (lack of rest, use of amphetamines, caffeine, etc.) while neglecting the necessary inhibition phases (relaxation, sleep, vacations, etc.). There are also many environmental circumstances as well as metabolic states that disturb a person's ability to rest, think, react appropriately, relax and sleep. For this reason, negative magnetic energy which evokes a normalization of inhibition is the most needed energy for humans. Therapeautic values of negative magnetic energy include relaxation, sleep and blocking of anxieties, phobias, obsessions and psychotic ideas.

Anxieties, phobias, obsessions, compulsions, depression, psychosis and seizures evoke an abnormal rise in the electrical activity of the CNS with an associated rise in positive magnetic energy; therefore, it is reasonable to assume that the CNS could be "tuned down", and thus controlled, through proper placement of negative magnetic energy. It has been noted that the higher the abnormal electrical activity of the brain, the higher the magnetic gauss strength needed to control the electrical activity. Correct magnetic gauss strength can cancel involuntarily evoked symptoms without interfering with voluntary initiation of thought, affect or motor responses. Actually CNS response to negative magnetic energy ranges from mild relief of symptoms (anxiety, stress, hyperactivity, abnormal thoughts) on one end of the scale to a state of general anesthesia on the other end of the scale. Such responses occur without harmful effect to metabolic function or tissues. (The magnets discussed herein are not strong enough to produce general anesthesia.)

Emotional and mental symptoms can be relieved by appropriately treating the brain with negative magnetic energy. Frontal, crown of head and back of head (occipital) placement of negative magnetic energy has a calming effect on the brain and spinal cord. However, it appears from clinical observation that bitemporal placement usually is most appropriate for emotional and mental illnesses.

Schizophrenics have been demonstrated to have chronic abnormally high electrical discharges from deep within the brain, especially from the amygdala. Post mortum studies reveal schizophrenics to have higher than normal amounts of the synaptic junction transmitter, dopamine, in the amygdala (mid-temporal area of the right and left hemispheres), with the highest amount being in the dominant hemisphere which would be the left hemisphere in the righthanded individual. Simultaneous bilateral placement of negative magnetic energy on right and left temporal areas provides maximum calming electrical control over

lymbic-hypocampus-amygdala and prefrontal areas of the brain. It is important to understand that the positive pole should not be used on the head for treatment since it is as exciting to brain function as the negative pole is calming to brain function. Emotional excitement, even to the extent of psychosis and seizures in patients so prediagnosed, can occur when positive magnetic energy is applied to the head.

Treatment goal is to calm down the electrical activity and excessive dopamine production occurring in the amygdala so as to reduce the evoked dissociation between the center of the brain and the prefrontal cortex. Progressive relaxation treatment should accompany negative magnetic energy treatment. Training out anxieties and phobias requires a reapproach to the stimuli evoking the symptoms. In treating obsessions and compulsions, the obsessional ideas should be reviewed while being inhibited by the temporal lobe magnetic energy treatment.

Some patients respond best using taped behavioral therapy drills while simultaneously placing 1-6 4X6" ceramic magnets at the crown of the head. The magnetic gauss strength required for symptom relief is in proportion to the degree of disordered electrical activity of the brain caused by the illness. Approximately fifteen minutes exposure time is required before a patient can communicate if thoughts and affect were out of control. When in control, responses can be initiated (such as relaxation), depression will be gone, euphoria if mildly present will be controlled while hallucinations, obsessional thoughts and compulsions will cease and the ability to voluntarily turn off and on thoughts and acts will emerge. At this stage of magnetic energy control of the brain a corrective reprogramming of the brain can take place. Essentially this is a process of unlearning. Maladaptive involuntary responses are inhibited and replaced with a new set of voluntary socially, mentally and physically appropriate responses.

In addition, some patients respond well by placing a 3X12" or 3X24" plastiform magnetic strip down the spine from one to several hours while proceeding with daily pursuits. This therapy has a quietening effect on the spinal reflexes and brain.

While a patient is being treated with magnetic facilitated behavioral modification for emotional &/or mental reactions, every effort should be made to stop the cause evoking the maladaptive behavior. Evoking factors can be food addictions, food allergies, hypersensitivity to chemicals, nutritional deficiencies, metabolic disorders or viral infections affecting immunity and brain function. Viral infections should always be considered seriously as a possible cause for maldaptive behavior. Diagnosis and treatment of organic factors in

emotional illnesses is the subject of Brain Allergies: PsychoNutrient Connection, William H. Philpott, M.D.

Negative magnetic therapy of the forehead can encourage relaxation, increase alertness and improve memory recall. The entire spine can be treated with a 3X24" or 4X24" plastiform magnet simultaneously. An effective treatment for tension, anxiety, phobias, depression, obsessionalism or psychotic ideas is simultaneous bitemporal placement. With the magnets placed over the temporal areas, the magnetic energy flows through temporal areas into the center of the brain where several nerve tracks cross. If placing the magnet on areas other than the temples reduces symptoms, use whatever placement is symptom relieving. Some patients find a 4X6" ceramic magnet on the back of the head provides more symptom relief than the bitemporal placement. Fifteen to 30 minutes is usually sufficient to stop symptoms. To stop symptoms, treatment may need to occur at intervals of 3-4 times daily. The relief of symptoms is directly proportional to the strength of the magnet. The stronger the symptoms, the stronger the magnet must be. Obsessional thoughts and psychotic ideas would require the strongest magnets such as the 2X5", 4X6" or 2X1 3/8" ceramic magnets.

As a behavioral therapist for 27 years, this medical author searched for 23 years for organic factors in major mental illnesses and spent l8 years systematically testing patients for genetic errors, nutritional disorders, infections, toxins and allergic-like reactions to foods, chemicals and inhalants (bio-ecologic medicine).

A positive double-blind study demonstrating a relationship between behavioral disorders and food reactions led to a systematic testing of mental patients' reactions to foods, chemicals and inhalants. The study compared the results of nontraditional bio-ecologic therapy with the results of traditional drug therapy in cases of major mental illness. When deemed valuable, patients in both groups received electric shock treatment. Over a three year period only 25% of the psychiatric patients receiving bio-ecologic treatment required a return to hospitalization while 60-75% of the patients receiving traditional drug therapy required rehospitalization. Obviously, allergic-like reactions to environmental substances, especially foods, are a significant factor in evoking mental illness. It seems equally obvious that schizophrenics may have an organic brain defect which would conceivably predispose them to less central nervous system stability in reactions to environmental substances such as foods, chemicals and inhalants.

Even though this author has demonstrated that allergic-like reactions to foods and chemicals can be prevented or reversed by magnetic therapy, it would be concluded that other factors should be examined and simultaneously treated for the most effective therapeutic results. (Refer to Brain Allergies: the Psycho-

Nutrient Connection).

A l7 year old girl presented with a diagnosis of schizo-affective illness and a history of depression with suicidal attempts, irritabilty, confusion and gastric upset. These major symptoms would last up to four days after eating gluten grains (wheat, rye, oats, barley). The girl was given a test meal of wheat. Headache, depression, irritabiltiy and gastric upset began to develop. For a 45 minute period she was exposed to negative magnetic polarity as follows:

l) 3X24" plastiform magnetic strip down spine
2) 4X6" ceramic magnet on epigastric area (stomach)
3) Ceramic cube magnet placed on each temple area

After 20 minutes of magnetic exposure, all symptoms left and she fell asleep. When she awoke, symptoms did not reemerge.

One day this same patient ate sole, a food which had not been allergy tested. She reacted with agitation, depression, negativism and social withdrawal. In this reactive and insightless state she refused to follow her mother's request to use the magnets for relief. She exercised vigorously but the exercise brought no relief. The next morning the symptoms continued. Again her mother suggested the use of the magnets. Grudgingly she accepted the suggestion. Within 30 minutes after application of negative magnetic energy to the temples, spine and sternum, the symptoms abated and did not return when the magnets were removed.

A 48 year old manic-depressive patient used the negative pole of a 3X24" plastiform strip magnet regularly on his spine to help control mood swings enabling him to maintain moderation in thought and behavior. In addition, he placed the cube magnets on the temples and the 3X24" magnetic strip on the spine for 30 minutes daily while practicing behavioral therapy tapes. The magnets allowed him to relax improving his ability to concentrate and practice the behavioral techniques.

A 37 year old paranoid man sat on a 3X8" plastiform magnetic strips for 30-60 minutes when overwhelming "feelings of fear" and head pressure bothered him. Usually complete relief was attained.

A young man with a long history of schizo-affective symptoms (depression & obsessionalism) requiring repeated medical intervention and hospitalization was able to inhibit obsessional thoughts by placement of magnets on the temples. There was a direct relationship between the strength of the obsessional thoughts occurring at any given time and the magnetic gauss strength needed to control the obsessional thoughts. As the obessional thoughts came under control the magnetic gauss strength could be reduced.Fifteen minutes of

treatment resulted in obsessional thought control for sometime after the removal of the magnets. When the obsesssional thoughts returned the magnets were replaced according to the strength needed for inhibition. The magnets were kept readily available for the patient to apply when symptoms began to develop. Negative magnetic energy controlled spontaneously occurring obsessional thoughts but did not interfere with voluntary initiation of thought, relaxation or motor acts and did not alter the state of consciousness. The ability to respond to a muscle relaxation drill normalized when magnets were on the temporal areas, whereas before, obsessional thoughts interfered with the patient's ability to relax and respond to verbal cues from the therapist.

A 44 year old man with anxiety and mania had difficulty slowing down thoughts and overactivity. By placing 3X12" plastiform magnetic strip on the lower spine for periods of time ranging from 30 minutes to several hours, he was able to normalize his energy drive--and even fall asleep.

A 58 year old woman with agitated depression and headaches could obtain relief in 5 minutes with a 3X4" plastiform magnetic strip applied to the left temporal area. She could fall asleep and when awakening the headache would be gone.

A 28 year old man, originally diagnosed as autistic at age 5, and as schizophrenic in adolescence exhibited numerous obsessions, compulsions and odd actions. He had been successfully controlled for ten years with a combination treatment program of ecology, nutrition and behavioral modification. He became careless with dietary habits and ate wheat more frequently than allowed whereupon he developed a compulsion to repeat the word "circle" over and over again. With placement of ceramic cube magnets on both temples the compulsion was relieved.

EMPHYSEMA

Using ceramic magnets or stacked plastiform magnets, apply magnetic therapy to chest/lung area and the lymph system. DO NOT USE MAGNETS ON CHEST WITH PACEMAKER OR DEFIBRILLATOR!

ENCEPHALITIS

Consult physician. Magnetic therapy may be given a trial using the CVS placement &/or temporal placement.

ENERGY, LOW

Low energy, which may be caused by a combination of factors, is best treated with full body therapy by sleeping on the Magnetic Bed nightly. During the day use magnets as described for Heart. Blood flow through the magnetic field over

the heart has been observed to increase biological energy.

If a Magnetic Bed is not available use Nightly Routine and during the day Heart instructions. Also refer to Maintaining & Increasing Energy in this Handbook.

EPILEPSY

Magnetic therapy has been observed to be very helpful in a number of seizure cases. Treatment must be done only under close medical supervison.

FATIGUE
Treatments for fatigue include l) total body treatment with Magnetic Bed or 2) Nightly Routine plus CVS placement. For further information refer to Sleep, Chronic Fatigue Syndrome, and Low Energy.

FEVER BLISTERS
Treat mouth/lips directly with ceramic magnets or plastiform magnetic strips 30-60 minutes 2 times daily. In addition, the Magnetic Bed or Nightly Routine could be helpful in inhibiting viral activity.

FIBROID TUMORS
Treat tumors directly with ceramic magnets or stacked plastiform magnetic strips for 60 minute intervals 2-3 times daily or all night.

FLATUS (gas)
Using ceramic magnets or stacked plastiform magnets, directly treat the distressed area--stomach, colon, etc.--until symptoms are relieved. If chronic flatus persists, consider food reactions and apply magnetic therapy as described in Food Reactions.

FLUID RETENTION See Edema.

FOOD REACTIONS
Symptom prevention: apply 15-30 minutes pre meal
Symptom relief: apply 15-30 minutes post meal

For minimal effectiveness:

a) HEAD: 3X6" plastiform strip magnet + round neodymium magnet placed on low occipital area (back of head/base of skull where skull joins soft neck tissue) Alternative: 4X6X1/2" flat ceramic magnet under head while reclining.
b) UPPER FOREHEAD at hairline 2X2" plastiform or 2X5" ceramic

For better effectiveness, add to the above:

c) STERNUM: 3X12" plastiform or 2X5" ceramic magnet
d) SPINE: 3X24" plastiform or

For best effectivness, add to above:

e) SPLEEN: 4X6" ceramic magnet (low back left side at junction of rib-age/abdomen)
f) LIVER: 4X6" ceramic magnet (low back right side at junction of rib-cage/abdomen)
g) STOMACH: 4X6" ceramic magnet

h) 4X6" ceramic magnet on areas of discomfort/swelling

Treatment of Foods

Some foods are less symptom producing if treated with negative magnetic energy before ingestion. To treat food, place in a glass container and set on the negative magnetic pole of a 4X6" magnet for a minimum of five minutes before eating.

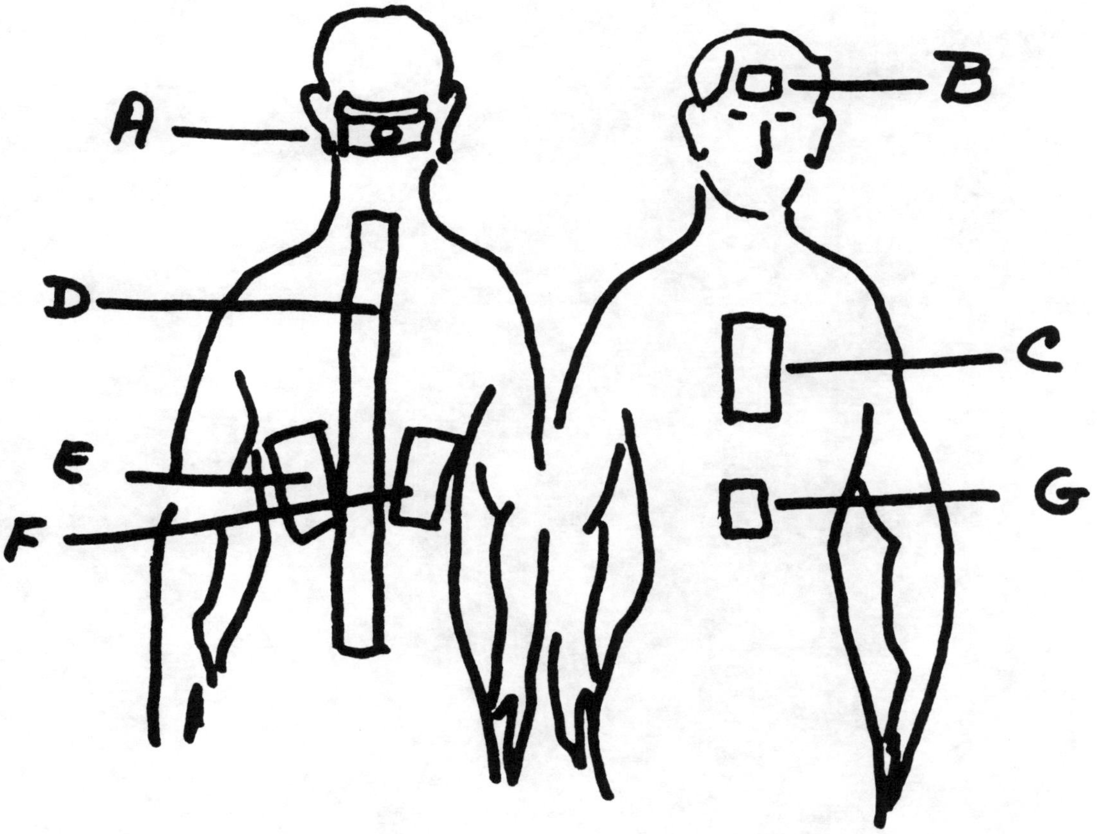

Following a test meal of buckwheat a 56 year old patient experienced vise-like head pressure, mental confusion, inability to respond appropriately when questioned, difficulty moving and pulse of 100 with skipped beats. A 4X6" ceramic magnet was placed under the back of the head. Within 5 minutes the patient had her eyes open, was mentally clear and could communicate easily and appropriately. The magnet was left in place for 30 minutes. Even after the

removal of the magnet improvment was maintained. Pulse slowed to 76 without skipped beats. The only complaint was a minor degree of residual head pressure.

A 14 year old boy complained of stomach upset, headache and irritability after eating a test meal of blackberries. He was treated with ceramic magnets on both temples, a magnetic strip on the abdomen and a magnetic strip down the spine for 30 minutes. He relaxed to the point of sleep and the headache and irritability improved.

FRACTURES, BONE

Robert O. Becker, M.D., orthopedist, has conducted detailed research into discovering the mysteries of bone healing--especially important in the instance of non-union of broken bones. Research demonstrates that a necessary ingredient in bone healing is a flow of electrical current to the injured site. Becker discovered electricity flows through the paraneural Schwann cells covering the nerves. If a nerve is severed, the Schwann cells quickly heal bridging the nerve-damaged area to provide conductivity of electrical current for the bone to heal. Healing of tissues other than bones proceeds in the same way demonstrating regeneration is always associated with an electrical current.

An intact bone is a weak magnet with both negative and positive magnetic fields present. When a bone is broken it becomes two separate magnets which repel each other. Then obviously, if broken bones are left to heal naturally, the fractured bone ends resist each other and will not come together smoothly. Thus, if fractured bones are misaligned, a cast must be applied to align the bone ends and hold in place until healing occurs.

When placing magnets over a fracture site, the magnetic energy field from the magnet overrides the weaker magnetic field of the bone--and the bone assumes the same magnetic field as the externally applied magnet. Thus by using magnets fractured bone ends are allowed to naturally come together without resistance. In addition, negative magnetic energy speeds healing by pulling oxygen and necessary nutrients for healing into injured tissues.

Calcium, also needed for bone healing, travels from normal bone tissues to injured sites. Flow of calcium can be facilitated by: l) Placing negative magnetic pole of a 4X6" ceramic magnet over the injured area while 2) placing positive magnetic pole of a 4X6" ceramic magnet 3-4 inches under the injured site facing the injury. A flow of energy exists between the positive and negative magnetic poles. With the push of the positive pole and the pull of the negative pole, calcium will move from the healthy bone area to the injured area speeding up bone knitting, if necessary. DO NOT USE POSITIVE MAGNETIC ENERGY WITH A PUNCTURE WOUND OR WHEN INFECTION IS SUSPECTED.

GASTROINTESTINAL DISORDERS & DISEASES

There are several types and degrees of inflammatory gastrointestinal diseases and disorders such as idiopathic inflammatory bowel disease, irritable bowel syndrome, regional enteritis (Crohn's Disease), colitis, ulcers (esophageal, gastric, duodenal), gluten enteropathy (allergy to gluten in cereal grains) and just plain chronic indigestion, "heartburn" and flatus. There is a common denominator in these disorders--usually food allergies.

EXAMPLES: Dairy products have been found to be especially notorious for causing ulcers. An allergy to gluten (protein found in cereal grains: wheat, rye, oats, barley) can create inflammation of the lining of the intestine causing a malabsorption syndrome (Celiac Disease) resulting in nutritional deficiencies, etc.

For most effective treatment, food allergies/sensitivities should be examined as described by this medical author in Brain Allergies: the PsychoNutrient Connection. Along with proper dietary changes, magnetic therapy can be useful for management of inflammatory gastrointestinal disorders. Use magnetic therapy l) premeal as described in Food Reactions to prevent reactions, 2) each evening as described in Nightly Routine and, 3) for specific symptom relief place 4X6" magnet over specific area of disorder. EXAMPLE: Regional eneritis usually occurs in the right lower quadrant of the abdomen which is where magnetic placement should occur. An ulcer should be treated directly over the involved area. Symptoms may be kept down by treating with magnetic therapy as long and as frequently as necessary to achieve comfort.

NOTE: The usual rule of thumb is DO NOT PLACE A MAGNET OVER THE ABDOMEN FOR 60-90 MINUTES POST MEAL which will allow sufficient time for peristalsis (wavelike muscular movement in digestive tract which moves food through) to occur. When a magnet is placed over the stomach or intestines, peristalsis is stopped. If gastrointestinal pain or other symptoms occur post meal, use magnets over the abdomen only long enough to relieve immediate symptoms and then remove allowing peristalsis to occur before reapplying.

GALLBLADDER

For gallbladder problems place 4X6" ceramic magnet directly over gallbladder as long and frequently as necessary or until symptoms are relieved. In addition, apply the Nightly Routine on a regular basis.

GEOMAGNETIC INFLUENCES, SYMPTOMS FROM

Refer to introductory section of this Handbook for further information. The Magnetic Bed would offer the most effective therapy to offset subtle influences from geomagnetic disturbances. In addition, specific symptoms should be treated as described in this Handbook.

GLAUCOMA

Glaucoma is a condition in which the pressure of the fluid in the eye is so abnormally high that it causes damage. Magnets should not be placed directly over the eye as this will draw fluid into the eye and increase pressure. Place ceramic magnet(s) or 1-3 stacked plastiform magnets to the side or below the eye to draw fluids away from the eye.

GOUT

Gout is a metabolic disorder that causes attacks of arthritis. Follow instructions for Arthritis.

HEAD

The positioning of magnets on the head varies with symptoms. If one placement does not achieve relief another placement should be tried until symptoms are relieved.

Bitemporal or temporal placement: magnets placed on both temples simultaneously or on one side at a time will help control affect, awareness and integration with other areas of brain cortex.

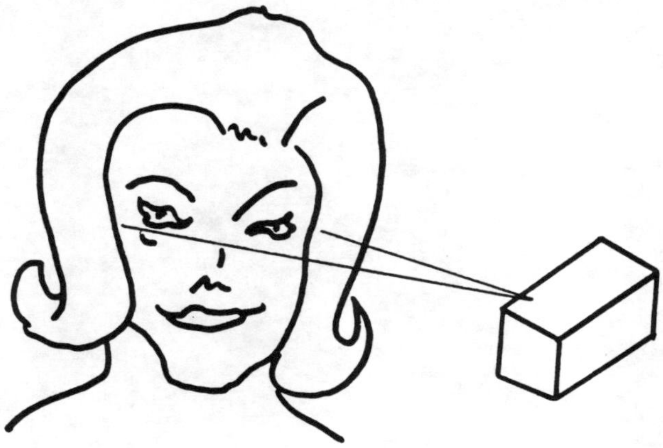

Cerebellar-vestibular placement: magnets are placed at the base of the skull so magnetic energy will radiate into base of brain for a calming effect. Also see Cerebellar Vestibular System.

Forehead placement: Any magnet(s) placed on the forehead will have a calming effect on emotion, will improve concentration capacity and help correct disordered thinking.

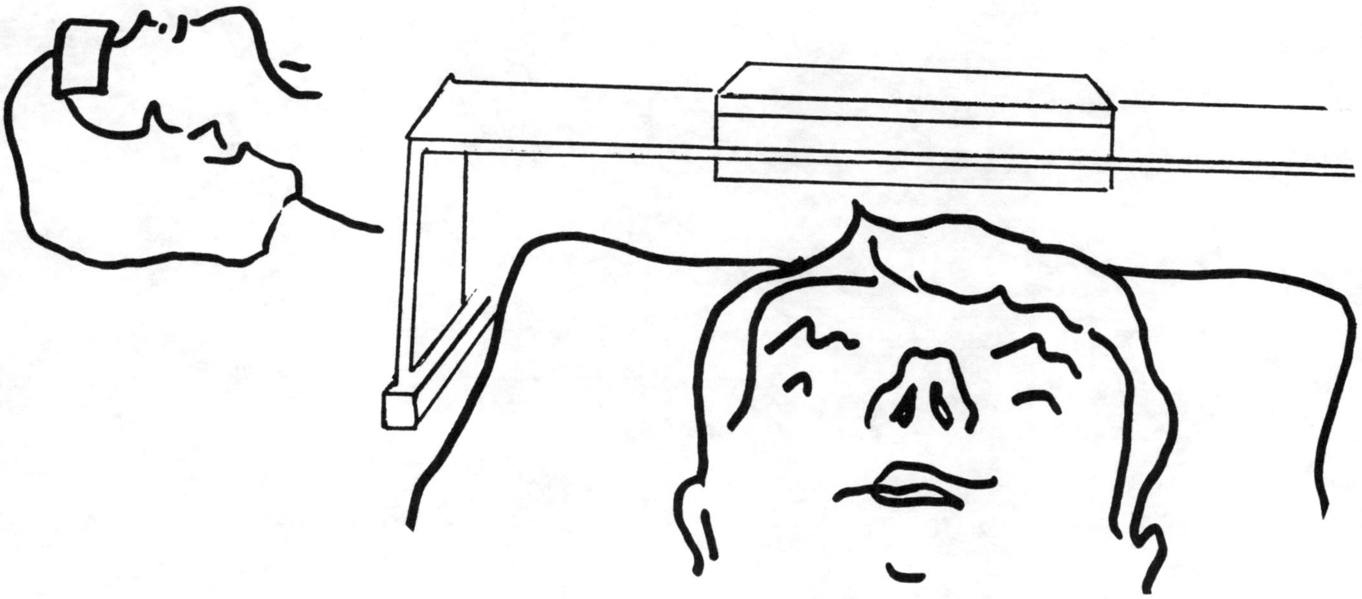

HEADACHE
Since headaches are caused by various conditions, trial placement of magnets may be necessary to find relief. Placements to try: l) Apply any ceramic magnets or stacked plastiform magnets directly over painful area. If no relief, place magnet(s) on opposite side of head. Fluid is pulled toward magnet(s) and depending on cause of pain, it may be necessary to pull fluid away from painful area. 2) Apply 2 ceramic cube magnets or 4X6" ceramic magnets bitemporally. 3) Place 1 plastiform strip plus l neodymium round magnet on low occipital area (back of head/base of skull) along with ceramic or plastiform magnet(s) on forehead at hairline. Refer to Head for illustration of suggested placements.

Food reactions and exposure to chemicals must be seriously considered as causes for headaches. Try to determine what caused the headache and avoid substances or stressful situations which may be contributory. For detailed information on testing foods and chemicals, see Brain Allergies: the PsychoNutrient Connection or contact this medical author.

HEALING, WOUNDS/SORES
Negative magnetic energy speeds healing of any wound, infected or noninfected, by pulling oxygen and necessary nutrients for healing into the injured tissues. In addition, it reduces inflammation and combats infection. Extensive research has demonstrated that a flow of electrical current to an injured site is necessary for healing. Robert O. Becker, orthopedist, discovered

that electricity flows through paraneural Schwann cells which cover the nerves. If a nerve is severed, the Schwann cells quickly heal bridging the nerve-damaged area providing conductivity of electrical current to the injured site for healing.

The affected area should be treated with direct application of ceramic magnets or stacked plastiform magnetic strips as continuously as possible until healing has occurred.

<u>HEARTBURN</u> See Gastrointestinal Disorders and Hiatal Hernia.

<u>HEART CONDITIONS</u>

WARNING! DO NOT USE A MAGNETIC BED OR MAGNETS OVER THE HEART WITH A PACEMAKER or DEFIBRILLATOR! Magnets within two inches of instrumentation can cause malfunction.

Magnetic therapy is helpful in normalizing cardiac irregularities such as tachycardia, palpitations and arrhythmia. Males can wear a 2X2" or 3X3" plastiform magnet over heart in shirt pocket; females wear a 2X2" plastiform magnet in left side of bra (no underwire) over the heart.

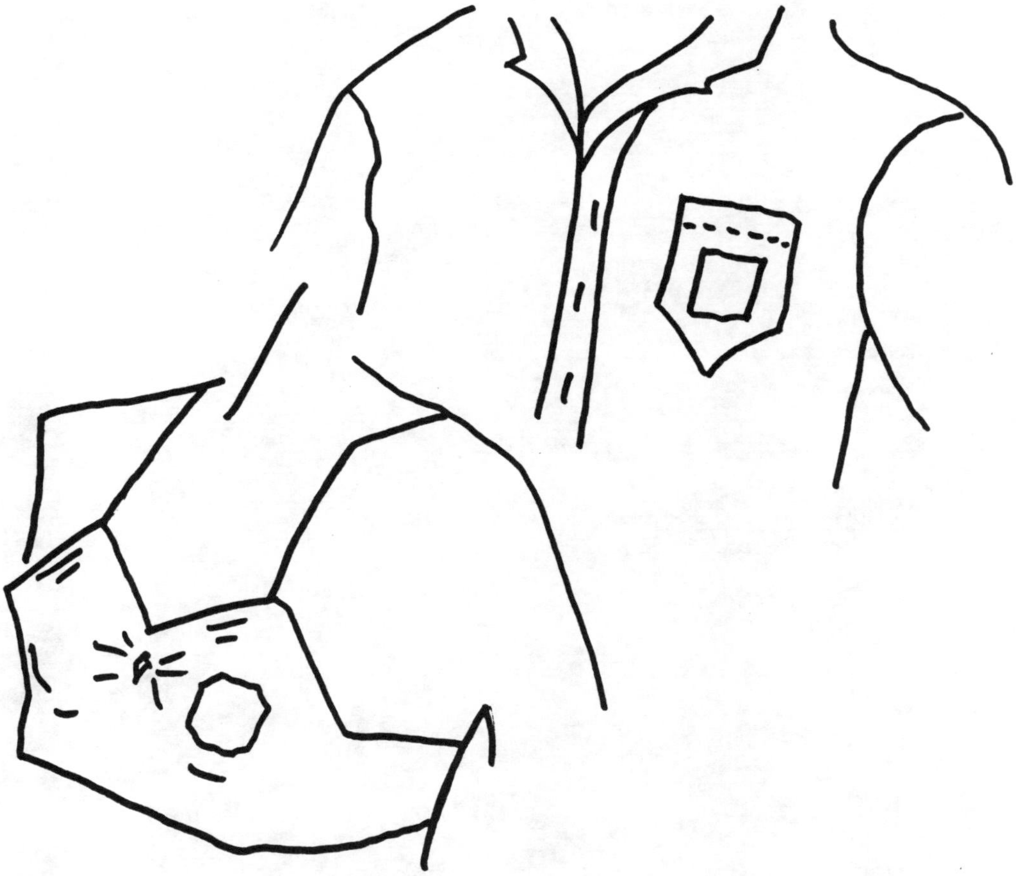

This placement of magnets also l) provides a systemic value to the entire body by oxygenating and conditioning (pH) the blood as it circulates through the heart. Has been demonstrated useful in treating mitral valve prolapse, high cholesterol and triglycerides, atherosclerosis as well as cardiac irregularities.

The magnets can be used as instructed l) all day while awake, 2) one or more times daily as preferred or 3) all night while sleeping.

In cases of atherosclerosis, 4X6" ceramic magnets should be applied directly to major areas of plaque. See Atherosclerosis.

HEMORRHOIDS

Colon and rectum should be treated directly by reclining or sitting on 4X6" ceramic magnet(s) or 3-4 stacked 4X12" plastiform magnetic strips. Magnetic energy will penetrate through the colon and abdomen.

HERNIA, HIATAL

A condition in which the stomach protrudes upward into the chest through a hiatus (opening) in the diaphragm often producing "heartburn". To alleviate symptoms, the sufferer should avoid eating large, heavy meals and should never lie down or bend over immediately after a meal. The head of the bed can be raised to prevent reflux during the night.

Magnetic energy applied by 4X6" ceramic magnet(s) to the stomach, diaphragm and painful area can be beneficial in relieving acidity and irritation.

HERPES, GENITAL

Herpes Simplex II, a viral infection, produces a painful rash in the genital area. After an incubation period of approximately one week, the virus produces itching, burning, soreness and small blisters. The blisters burst to leave small painful ulcers which heal within 10-21 days. Lymph nodes in the groin may become enlarged and painful with accompanying headache and/or fever.

A Magnetic Bed provides optimum therapy for viral infection and for prevention of infectious flareups. For relief and healing of acute genital herpes flareup, apply 4X6" ceramic magnet(s) to the infected genital area as well as the sacro-lumbar spine (low back area) where infected neurons are located. Duration of treatment: 10-21 days or more as continuously as possible. It would be advisable to also treat involved lymph nodes and the spleen with 4X6" ceramic magnet(s). Followup treatment should consist of 30-60 minutes exposure twice daily for several more weeks.

For further information refer to Infections, Viral.

This placement [illegible] is used [illegible] for [illegible] (and [illegible]) [illegible]. This [illegible] has been [illegible] useful [illegible] and [illegible], as well as [illegible].

[illegible] can be used [illegible] during the day [illegible] awake [illegible] as [illegible].

[illegible] of [illegible] magnets should be applied directly [illegible] major areas of pain. See Mineral [illegible].

HEMORRHOIDS

Dilated [illegible] veins [illegible] (anal) [illegible] with [illegible].

HIATAL HERNIA

A condition in which the stomach [illegible] has moved into the chest through [illegible] opening in the diaphragm, often [illegible]. [illegible] symptoms the sufferer should avoid [illegible] lying down, or bend over immediately after a meal. The head [illegible] raised to prevent reflux during the night. [illegible]

[illegible] can be [illegible] and [illegible] area can be [illegible].

HERPES SIMPLEX

Herpes simplex I, a viral infection [illegible]. After [illegible] approximately one week [illegible] tingling, burning [illegible] and [illegible] painful [illegible]. [illegible] become [illegible].

[illegible] Magnet [illegible] of [illegible]. For [illegible] apply [illegible] magnet(s) to [illegible] area [illegible] lesions [illegible] possible [illegible] also [illegible] immune [illegible] exposure [illegible] several more weeks.

[illegible] infection.

HICCUPS

A sudden involuntary contraction of the diaphragm followed by rapid closure of the vocal cords which causes the characteristic sound. Common hiccups occurring for brief intervals and lasting only a short time may be caused by food reactions.

For prolonged irritating hiccups, apply 4X6" ceramic magnet(s) or stacked plastiform magnets to the diaphragm and stomach area until symptoms are relieved.

HOT FLASHES

Since "hot flashes" are usually caused by decreased ovarian hormone production, the reproductive system (particularly ovaries) should be treated directly with 4X6" ceramic magnet(s) or 3-4 stacked plastiform magnetic strips. In addition the Nightly Routine should be used regularly.

HYPERACTIVITY

Hyperactivity is characterized by overactivity and inability of the individual to concentrate on one activity very long. Probably the most practical magnetic application for a child would be the Magnetic Bed since it would require no voluntary cooperation on the part of the individual. In addition, the Sleep application should be used during sleeping hours in conjunction with the Magnetic Bed. For symptoms during waking hours, trial should be made using CVS and other head placements until symptomatic relief is achieved.

If a Magnetic Bed is not available, negative magnetic energy should be applied with ceramic magnets or stacked plastiform magnets on each temple a minimum of 30-60 minutes 1-3 times daily. In addition, trial placement should be made as described in CVS. The Nightly Routine should be used on a regular basis.
Food allergies, chemical exposures and metabolic disorders should be considered as causes for hyperactivity. Hyperactivity, if untreated in children, often leads to schizophrenia in adults. If the cause of hyperactivity is diagnosed early in life and treated properly, most usually the illness can be reversed. Refer to Brain Allergies: the PsychoNutrient Connection, by William H. Philpott, M.D.

HYPOGLYCEMIA (Low blood sugar)

It has been observed by this medical author that low blood sugar, traditionally believed to be solely a carbohydrate intolerance problem, may be evoked by foods of all types whether fats, carbohydrates or proteins. Foods causing hypoglycemia are specific in each individual. The central problem in hypoglycemia and diabetes is not, therefore, that of a general type food (i.e., carbohydrate), but that of an individual maladaptive reaction to a specific food or chemical which in turn interferes with proper blood sugar levels.

In addition to testing foods and making dietary adjustments as described in Victory Over Diabetes: a BioEcologic Triumph, magnetic treatment pre and/or post meal as described in Food Reactions will provide a degree of relief.

HYPERTENSION

Hypertension, or high blood pressure, is a common condition affecting 10% of the adults in the U.S. It is diagnosed when the resting blood pressure (diastolic) is persistently raised. Although hypertension rarely causes symptoms, left untreated it increases risk of stroke and other disorders.

Hypertension can be aggravated by emotional stress, food reactions, chemical exposures, toxins, kidney infections, etc. Since causes for hypertension vary, magnetic treatment varies from patient to patient.

Ideal treatment would involve full body treatment with a Magnetic Bed which would relieve stress, treat blood flowing throughout the body, promote processing of toxins and dissolution of plaque buildup in arteries. If a Magnetic Bed is not available, 4X6" ceramic magnet(s) or 3-4 stacked plastiform magnetic strips should be placed on the kidneys 30-60 minutes 2-3 times daily. In addition, the Nightly Routine should be used regularly.

HYPOTHYROIDISM

Hypothyroidism is underactivity of the thyroid gland and, therefore, underproduction of thyroid hormones which stimulate energy. Classic symptoms are generalized tiredness and lethargy. There may also be muscle weakness, cramps, slow heart rate, hair loss and deepening of the voice. Hypothyroidism may be due to l) inhibition of the thyroid gland or 2) damage of the thyroid gland due to a viral infection or autoimmune disorder. Therefore, the degree of effectiveness in treating hypothyroidism varies.

For most effective thyroid therapy, rotate ceramic magnet(s) or 3-4 stacked plastiform magnets from the 7th cervical (back of neck) to throat/thyroid area (front of neck). Treatment should be 30-60 minutes 2 times daily. In addition, instructions for Sleep should be followed regularly.

HYPOTENSION

Since some healthy people have normally low blood pressure, the term, hypotension, is usually only used when blood pressure has fallen to the extent that blood flow to the brain is reduced causing dizziness and fainting. Symptoms often occur when sitting up or standing up. Treatment depends on the underlying cause such as adverse effect from antidepressive or antihypertensive drugs, diabetes or physiological shock.

Magnetic therapy should include direct treatment of the adrenal gland and thyroid gland. For most effective thyroid therapy, rotate ceramic magnet(s) or 3-4 stacked plastiform magnets from the 7th cervical (back of neck) to throat/thyroid area (front of neck). Treatment for Cerebellar Vestibular System should followed. In addition, Sleep instructions should be followed daily.

INDIGESTION

Indigeston covers a variety of symptoms such as heartburn, abdominal pain, nausea, flatus (gas) in stomach or intestine causing belching and discomfort. Indigestion can be caused by food allergies, spicy/fatty foods and by eating too much too quickly. Nervous indigestion is a common effect of stress.

Self-help treatment includes avoiding foods and situations that evoke symptoms or stress and finding time to eat at regular intervals without rushing. Magnetic therapy as described in Food Reactions is applicable. For minimum treatment, place 4X6" ceramic magnet(s) or 3-4 stacked plastiform magnets over the distressed abdominal area until symptoms are relieved.

INFECTIONS, BACTERIAL

Minor skin infections: Apply one or more plastiform magnetic strip(s) to infected area continuously 7 days or longer if needed. The neodymium is ideal for areas l" or smaller.

Lung infections: Apply 4X6" ceramic magnet(s) 3-4" apart over affected area as continuously as possible 4-7 days or longer. Lungs should be treated with magnetic energy from the front and back simultaneously. Occasionally prolonged treatment cannot be tolerated due to bronchial constriction from negative pole magnetic exposure. In such case, treat up to level of tolerance but treat as frequently and as long as tolerated. DO NOT USE ON CHEST AREA WITH PACEMAKER OR DEFIBRILLATOR!

Urinary bladder & kidney infections: Apply 4X6" ceramic magnet(s) as continuously as possible to affected area 2-3 days beyond symptom clearance.

Antibiotics: Antibiotics can be combined with magnetic treatment but will likely not be needed if negative magnetic field energy is used with sufficient gauss strength, appropriate placement and adequate duration. The more areas treated, the longer the treatment time and the higher the gauss strength, the more successful the treatment will be and less likely the need for antibiotics. If antibiotics are needed, magnetic energy will enhance the effect of the medication.

INFECTIONS, FUNGAL

Candida albicans (yeast infection) is a fungal organism which is present in all humans but does not pose a problem unless it multiplies to the extent that the immune system can no longer control it. Candida is an opportunist organism which thrives when body defenses are "down" for some reason. The most common causes for low body defense include l) chronic viral infection 2) nutritional deficiencies, 3) metabolic shifts for various reasons and/or 4) use of

broad spectrum antibiotics which will kill "good bacteria" along with bad bacteria. "Good bacteria" works to keep candida under control and if destroyed, candida takes the opportunity to multiply and establish itself. Symptoms may be physical as well as mental and most frequently are manifested with weakness, poor concentration, confusion, fatigue, depression, difficulty making decisions, plus others. In addition, candida manufactures toxins throwing them off into the body with the end result that often food and chemical allergies develop.

Yeast infection can be systemic (throughout body systems) or may be localized in nature. Examples of localized infection would include athletes foot, vaginal yeast infection, oral thrush, nail (toe & finger) infections, etc. Candida may be in the colon, lungs, mouth or even on the skin.

Treatment for candida infections involving lower abdominal area (genital area, vagina, bladder, intestinal tract, etc.): Sit on the negative pole of 1-3 4X6" ceramic magnet(s) 1-2" apart OR 2-3 stacked 4X12" plastiform magnets placed on chair seat. Treat a minimum of 3 hours daily. When sitting on the magnets, energy will radiate approximately as far up as the umbilicus thus treating the vagina, urethra, urinary bladder and lower colon.

NOTE: Magnets should be used premeal and later in the day. It is not advisable to sit on magnets for at least 60 minutes post meal since theoretically it is possible that negative magnetic energy could slow peristolsis.

Many candida (yeast) cases have been treated successfully with magnetic therapy by this medical author. One case in particular involved an obese woman with frequent fluctuating urinary bladder infections and/or nonspecific inflammation and vaginal candidiasis who had been treated repeatedly and unsuccessfully with antibiotic therapy. After sitting on negative magnetic energy for one hour intervals 1-3 times a day no further evidence of either bacterial or fungal infection occurred.

For oral infections (mouth/throat), 4X6" ceramic magnet(s) are placed on either side of the face and neck 30-60 minutes daily.

Athlete's foot can be treated by strapping the negative pole of a plastiform magnetic strip to the soles of the feet for at least 1-3 hour(s) a day. Even better, wear a magnetic insole in the shoe. Insole may be cut from plastiform material with knife and placed in shoe or may be purchased.

All microorganisms (fungal, bacterial, viral) and parasites are activated by positive magnetic energy and are stunned or retarded by negative magnetic

energy. Two mechanisms are likely at work producing this effect. l. Negative magnetic energy alkalinizes (normalizes pH) tissues. Microorganisms and parasites flourish in acid media but are inhibited in alkaline media. 2. Negative polarity increases cellular oxygenation. Evidence indicates that the counter-lockwise rotation of negative magnetic energy pulls oxygen toward it, in this case into the cell. Oxygen is needed for human cell metabolism but stuns microorganisms and parasites which have a low tolerance for oxygen.

INFECTIONS, VIRAL

Epstein-Barr, cytomegalovirus and Human Herpes-6 (HHV-6) are herpes-family viruses which cause infectious mononucleosis. Classic symptoms include extreme fatigue and lethargy, depression, and often flu-like symptoms. Immune system tissues and brain neurons are both particularly susceptible to these viruses. Thus in chronic viral cases patients usually present allergies and hypersensitivities to a large number of substances (immune system response) and suffer from extreme mental fatigue and depression (brain infection).

Optimum treatment would involve full body treatment with a Magnetic Bed. Negative magnetic energy inhibits the replication of viruses.

If a Magnetic Bed is not available, 4X6" ceramic magnet(s) or 3-4 stacked plastiform magnetic strips should be placed on the tissues involved such as head, lymph nodes, spleen and thymus as continuously as feasible day and night for 5-7 days or more. The Nightly Routine may be followed as treatment protocol. Sleep instructions should be observed regularly using Method I which treats the brain for infection. If enlarged lymph nodes are observed, affected tissues should be treated with ceramic magnet(s) or 3-4 stacked plastiform magnets for 30-60 minutes 1-2 times a day until swelling subsides.

Also refer to Shingles (Herpes Zoster).

INFLUENZA

The Magnetic Bed has been repeatedly observed to relieve discomforts of flu to a great degree, shorten the duration of the illness and avoid ongoing complications such as prolonged cough, lethargy, pneumonia, etc. If Magnetic Bed is not available, treat as for Infections, Viral.

During a flu epidemic drink magnetically treated water every 4 hours for prevention or relief of symptoms:

Magnetically Treated Water

With the drinking of magnetically (negative) treated water, the beneficial

effects of negative magnetic energy are passed on to the biological system. To treat water:

I) Place water in glass container
2) Set container on negative pole of a magnet

OR

Tape magnets to container with negative pole directed towards contents
3) Treat for minimum of 5 minutes, preferrably longer

A husband and wife in their 60s developed influenza with aching muscles and multiple body pains. The wife started out with a sore throat. She underwent no treatment for symptoms and developed full blown influenza and viral pneumonia requiring extensive medical treatment. The husband was with her during the entire illness sleeping in the same bed. With the first appearance of symptoms (swelling lymph nodes), he immediately started negative magnetic therapy applied to swollen lymph nodes and thymus gland 60 minutes each evening and 30 minutes at noon. For aching muscles, magnetic strips were applied to local tissues. Treatment continued through the duration of his wife's illness. His aching muscles disappeared, swollen lymph nodes receded and full-blown influenza never developed.

While it would be statistically impractical to generalize from a single case, this type of situation has been observed repeatedly. It appears logical to conclude a full-blown case of influenza can be averted by prompt application of negative magnetic field energy when symptoms first appear.

INSOMNIA

The Magnetic Bed offers optimum potential for stress reduction, muscle relaxation and pain relief which most often interfere with good sleep.

If the Magnetic Bed is not available, follow Nightly Routine or Sleep instructions. A 2X5" ceramic magnet should always be placed on the sternum providing a calming effect on lungs and heart. A feedback message to the brain indicates that all is well with the heart/lung area which encourages sleep. NOTE: DO NOT USE MAGNETS ON CHEST WITH PACEMAKER OR DEFIBRILLATOR!

In addition any stimulating substances such as caffeine should be avoided. Food allergies and chemical sensitivities can also evoke insomnia. See Brain Allergies: the PsychoNutrient Connection, William H. Philpott, M.D.

JET LAG

For prevention and/or jet lag recovery, treat the pineal gland as described in Sleep. The pineal gland should be treated 3-5 days ahead of travel and 3-5 days following travel. Additionally, adverse symptoms may be avoided by using CVS treatment during travel.

Other self-help suggestions: l) avoid the use of caffeine, 2) drink extra water and 3) cut back on food intake during travel.

LARYNGITIS

Laryngitis, inflammation of the voice box, is usually caused by infection and results in hoarseness. Magnetic therapy using ceramic magnet(s) or 3-4 stacked plastiform magnetic strips should be directly applied to throat, chest/lung area as continuously as possible until symptoms begin to subside. In addition lymph nodes in the neck need to be treated. Refer to Lymph Nodes & Carotid Arteries. Cerebellar Vestibular protocol may also be helpful. NOTE: DO NOT USE MAGNETS ON CHEST WITH PACEMAKER OR DEFIBRILLATOR!

LEARNING DISORDERS

Learning disorders, attention deficits, dyslexia, etc., often are also associated with hyperactivity. Metabolic disorders and food allergies should seriously be considered as contributing factors.

Negative magnetic energy should be applied with ceramic magnets or 3-4 stacked plastiform magnets on each temple for 30-60 minutes 1-3 times daily. In addition, trial placement should be made as described in Cerebellar Vestibular treatment protocol. The Sleep protocol should be used nightly.

LIVER TOXICITY

Apply negative magnetic energy directly to the liver (right side of abdomen below rib cage) using 4X6" ceramic magnet(s) for 30-60 minutes twice a daily, or oftener as needed. Treat liver from back of the body. If symptoms are not relieved, try treating from front of body.

LUPUS ERYTHEMATOSUS

Lupus, an inflammatory disease of connective tissue, can affect many systems of the body including joints and kidneys. Gluten enteropathy (see Gastrointestinal Disorders) presents many of the classic symptoms of lupus and can easily be mistakenly diagnosed as lupus erythematosus. When lupus symptoms are present food allergies and possible gluten enteropathy should be seriously considered and examined.

Ideal magnetic treatment would involve use of a Magnetic Bed providing full body therapy with general systemic value.

If a Magnetic Bed is not available, Nightly Routine should be followed regularly.

LYMPH NODES

Lymph nodes and carotid arteries are treated simultaneously by placing 2 ceramic cube magnets on each side of the neck anteriorly. 2X5" ceramic magnets or 3-4 stacked plastiform magnetic strips may also be used and held in place by a 4" Ace bandage.

Educational only. Not to be construed as prescribed medical treatment for any individual. If you have a health problem, consult your physician.

Treat the entire lymph system as follows:

1) Lymph nodes as directed above
2) Spleen: 4X6" ceramic magnet(s) applied 30-60 minutes twice daily or more if needed. The spleen is on the left side of the body at the junction of the ribcage and abdomen
3) Thymus: Neodymium magnet, 2X5" ceramic magnet or 3X6" plastiform magnetic strip applied over sternum 30-60 minutes twice daily or more if needed.

MACULAR DEGENERATION
Macular degeneration, often a complication of diabetes mellitus, should be treated as for Diabetes. Magnetic therapy may be applied directly over the eye with ceramic magnet(s) or 3-4 stacked magnets a minimum of 3 hours daily. Statistics have not been gathered on the effectiveness of magnetic therapy in macular degeneration.

MAGNETIC DEFICIENCY SYNDROME
Optimum magnetic therapy for Magnetic Deficiency Syndrome would be full body treatment with a Magnetic Bed which would provide a field of restorative negative magnetic energy about the body for relaxation, stress reduction and sleep promotion which all help restore the energy balance of the entire body.

If a Magnetic Bed is not available, the pineal gland should be treated regularly as described in Method II of Sleep. In addition, ceramic magnet(s) or 3-4 stacked plastiform magnets should be applied directly to areas of discomfort a minimum of 3 hours daily or until relief is achieved.

MENOPAUSAL DISCOMFORTS
Many of the discomforts of the menopausal phase (hot flashes, night sweats, insomnia, poor memory/concentration, tearfulness, depression and anxiety) can be relieved with magnetic energy.

Ideal magnetic treatment would involve use of a Magnetic Bed
which would provide full body therapy and general systemic value.

If a Magnetic Bed is not available, the Nightly Routine should be followed regularly. Additionally, reproductive glands (ovaries) should be treated directly with ceramic magnet(s) or 3-4 stacked plastiform magnetic strips 30-60 minutes 1-3 times daily or when hot flashes or other symptoms occur.

Additionally, symptoms of depression, anxiety, etc., can be treated per instructions in this handbook for each specific condition.

MENSTRUAL IRREGULARITY Melatonin, produced by the pineal gland, is deficient in most menstrual irregularities. Melatonin may be raised by placing the negative magnetic field at the crown of the head by either Method I or II described in Sleep.

MENTAL DISORDERS See Emotional & Mental Disorders

MENTAL FOGINESS ("Cobwebs" in brain)
Place 2X5" ceramic magnet or plastiform magnetic strip on forehead to

encourage relaxation, increase alertness and improve memory recall. Treatment protocol for Sleep used nightly can also increase alertness.

MIGRAINE HEADACHE

Oftentime migraine headaches are caused by food or chemical allergies/hypersensitivities. Examination should be made to determine, if possible, the cause of headaches.

For pain relief, try different magnetic treatment placements for the Head. If one placement does not bring pain relief, try other placements until symptoms are relieved.

MITRAL VALVE PROLAPSE See Heart.

MOTION SICKNESS (Land, Sea, Air)

Try different placements until comfort is achieved. Treat the ears with neodymium magnet(s). Place 2X5" ceramic magnet or plastiform magnetic strips on back of neck. Try plastiform magnetic strip(s) down spine. If comfort is not achieved, combine all three treatments.

MUSCLE SPASMS

Muscle spasms may be treated by placing ceramic magnet(s) or 3-4 stacked plastiform magnets directly over the painful area. Leg cramps may be relieved by placing ceramic magnets on the soles of the feet.

NAUSEA

Nausea resulting from flu, overindulgence in food or alcohol, etc., may be treated by using ceramic magnet(s) or 3-4 stacked plastiform magnets over the stomach, liver and gallbladder for 30-60 minutes or longer until nausea subsides. If nausea persists, contact a physician.

NERVOUSNESS

For a "case of nerves" try different magnetic placements: 3X24" plastiform magnetic strip on the spine, the Cerebellar Vestibular protocol, plastiform magnetic strip(s) on forehead or plastiform strips on temples. Treat until nerves are calm.

NEURITIS

Painful inflammation of a nerve, neuritis, can be caused by viral infection such as herpes zoster. Magnetic therapy should include treatment of: l) spine with 3X24" plastiform magnetic strip,
2) Cerebellar Vestibular System and 3) painful area with ceramic magnet(s) or 3-4 stacked plastiform magnetic strips.

NIGHTLY ROUTINE

Thirty to sixty minutes before retiring for the night, apply negative magnetic energy:

a) HEAD: 3X6" plastiform magnet + round neodymium magnet placed on low occipital area (back of head/base of skull where skull joints soft neck tissue) Alternative: 4X6" ceramic magnet under head while reclining.
b) SPLEEN: 4X6" flat ceramic magnet (L side low back at junction of ribcage/ abdomen
c) LIVER: 4X6" flat ceramic magnet (R side low back junction of ribcage / abdomen)
d) THYMUS: 2X5" ceramic or 3X6" plastiform magnet (anterior mid chest)
e) NECK LYMPH NODES: 2X3" plastiform magnets over lymph nodes on upper anterior aspect of neck under chin. Alternative: 2 cube magnets one on each side of neck or l cube magnet on one side + (2) 2X2" plastiform magnets stacked on opposite side.

Upon retiring for the night: Negative magnetic energy should be applied at the crown of the head all night during sleep using 4 4X6" ceramic magnets standing upright l" apart. The crown of the head (top) should rest 2-3" from the magnets during sleep placing the head in the negative magnetic field all night. Alternative: Apply negative magnetic energy by a round Neodymium to the crown of the head (exact center top of head) all night. Not only will this enhance deep sleep with physical and mental energy but will treat the pineal gland responsible for production of melatonin which controls the cyclic pattern of sleep and wakefulness.

OBESITY (Spot reducing)
Negative magnetic energy is alkalinizing and, therefore, can neutralize and dissolve fatty deposits. Local fat can be removed by sufficient magnetic strength and duration locally applied.

Abdominal fat: 3 4X12" plastiform magnets stacked together over the upper or lower abdomen as desired. Use all night held in place with abdominal binder. Always end up the last two hours with magnets on the lower abdomen to prevent constipation. Thighs: 4X6" plastiform magnets applied to fatty area
Buttox: sit or recline on ceramic or plastiform magnet(s)

The longer the exposure time the better. Spot reducing with magnetic energy is a very slow process. Also see Weight Reduction.

Fatty tumors have also been dissolved with direct application of negative magnetic energy using 4X6" ceramic magnet(s).

OBSESSIVE-COMPULSIVE DISORDERS
For detailed information refer to Emotional & Mental Disorders. Therapy should always include treatment of the Cerebellar Vestibular System.

OSTEOPOROSIS
Osteoporosis, decreased bone density, is a natural process of aging. Causes may be from removal of ovaries, calcium deficiency, certain hormonal disorders or long term corticosteroid treatment.

Ideal magnetic treatment would involve use of a Magnetic Bed
which would provide full body therapy with general systemic value. Additionally, affected areas should be directly treated with ceramic magnet(s) or 3-4 stacked plastiform magnets 30-60 minutes 3 times daily.

If a Magnetic Bed is not available, the 7th cervical vertebra should be treated with ceramic magnet(s) or 3-4 stacked plastiform magnets 30-60 mintues 3 times daily. In addition, negative magnetic energy with either type of magnets should be applied directly over the affected areas.

OVARIAN CYSTS
Ovaries may be treated directly with ceramic magnet(s) or 3-4 stacked plastiform magnets. Ideally, using an Ace Bandage or other abdominal binder, magnets can be bound over ovaries at night during sleeping hours. If discomfort from the cysts is present, 1-3 plastiform strip magnets worn during the day may provide additional relief.

Do not treat during menstrual period since negative magnetic energy may stop the period.

PAIN, CHRONIC
For pain relief refer to specific condition, e.g., Headache, Arthritis, etc.

Since pain can be caused by different conditions, the CAUSE of pain should be determined by self-test. l) Pain can be caused by a pressure from fluid buildup within a joint or other encapsulated area. Since negative magnetic energy pulls additional fluid into the treated area, pain can increase if direct application is made to a joint already full of fluid. In this instance, moving the magnet above, below or aside the painful site 2-3" will pull fluid from the affected area and may relieve pain. 2) Pain can also be caused by inflammation of a joint or other area. In this instance, direct application of magnetic therapy may provide relief by reducing swelling and inflammation.

For miscellaneous localized pain, trial application of any magnet(s) over the affected area should be made. When the appropriate application has been determined, treatment should continue with simultaneous treatment of the Cerebellar Vestibular System. Treatment may be 15-60 minutes or until symptoms are relieved and may be repeated as needed for pain.

For generalized chronic pain, ideal magnetic treatment would be a Magnetic Bed which would provide full body therapy and general systemic value. If a Magnetic Bed is not available, the Nightly Routine may be followed regularly.

A 49 year old patient experienced a severe headache after an allergy test for walnuts. She was treated with a magnetic strip down the spine and ceramic magnets on the right and left sides of her head. Within 30 minutes she was relaxed and the headache was relieved.

A 38 year old woman was suffering pain in the right lower abdominal quadrant with jerking muscles of the right leg. Upon examination the abdominal area was tender to palpation and she described a recent hysterectomy at which time the right ovary was observed to be enlarged. Application of negative magnetic energy with a 4X6" ceramic magnet relieved all symptoms including the jerking leg muscles.

PANCREAS
Apply 4X6" ceramic magnet(s) 30-60 hours twice daily or more if needed to the pancreas--directly over the epigastric area. Treatment should be 1 1/2 to 2 hours after eating in order not to inhibit insulin production during and immediately after meals.

PANIC ATTACKS

Panic attacks, anxiety and depression are all likely to be helped by raising melatonin by sleeping with negative magnetic energy at the top of the head (See Pineal Gland and Sleep, Method II). In addition, trial placements of magnets should be used to achieve maximum relief such as Bitemporal Placement, CVS, and Nightly Routine. Also refer to Emotional Disorders.

PHOBIAS See Emotional & Mental Disorders.

PLEURISY

Pleurisy causes a sharp chest pain that sometimes travels to the tip of the shoulder on the involved side. Magnetic therapy should be applied with ceramic magnet(s) or 3-4 stacked plastiform magnets to the direct area of pain. Treatment should continue 30-60 minutes 3 times daily or until relief is obtained.

PNEUMONIA

Consult physician if pneumonia is suspected. DO NOT USE MAGNETS ON CHEST WITH PACEMAKER OR DEFIBRILLATOR!

Optimum magnetic therapy would involve full body treatment with a Magnetic Bed PLUS Sleep Method I. Treatment should be as continuous as possible 5-7 days or more or until symptoms subside. Continuing treatment would involve sleeping on the Magnetic Bed and using 4X6" ceramic magnets at crown of head nightly.

If a Magnetic Bed is not available, ceramic magnet(s) or 3-4 stacked plastiform magnets should be placed on tissues involved such as chest/lung area and lymph system (lymph nodes, spleen & thymus). The Cerebellar Vestibular treatment protocol should be applied also.

PMS (Premenstrual Syndrome)

The Magnetic Bed offers optimum potential for relieving PMS symptoms. If a Magnetic Bed is not available, the Nightly Routine should be used as often as needed to relieve symptoms. Treatment should continue on a regular basis to prevent development of symptoms. In addition, the ovaries may be treated with ceramic magnet(s) or 3-4 stacked plastiform magnets during times of discomfort and 30-60 minutes daily between menstrual periods.

For specific symptoms associated with PMS, refer to the condition; e.g., for headache, refer to Headache herein.

PREGNANCY

DO NOT use magnets on abdomen during pregnancy.

PREVENTIVE THERAPY, GENERAL

For preventive measures and/or treatment of symptoms and diseases total body treatment with a Magnetic Bed provides the best overall benefit. The Magnetic Bed places the entire body in a negative magnetic field--Nature's own health restoring environment which is sleep inducing, stress reducing and anti-inflammatory. The resulting benefits of good sleep and stress management cannot be overemphasized. Although total body treatment is preferrable, if a magnetic bed is not available, the following Nightly Routine may be observed.

RUPTURED DISK

The painful inflammed area of rupture should be treated directly with ceramic magnet(s) or 3-4 stacked plastiform magnets. Negative magnetic energy reduces inflammation and swelling of tissues and thus reduces irritation and pain.

SCAR TISSUE

Scar tissue forms due to insufficient oxygen at a wound site during the healing process. It has a positive magnetic effect in the body and can interfere with energy flow and energy levels. No matter how scar tissue is formed (inflammatory reaction, surgery, etc.), it will reduce in size and may even disappear under the influence of negative magnetic field energy. Internal and external scarring should be treated directly with ceramic magnet(s) or 3-4 stacked plastiform magnetic strips as often as possible or all night until symptoms or scars disappear.

SCHIZOPHRENIA See Emotional & Mental Disorders.

SCIATICA

Treat the spine with 3X24" plastiform magnetic strip. Additionally, the lower back may be treated with ceramic magnet(s) for pain relief.

SEIZURES

Multiple case histories indicate the effectiveness of negative magnetic therapy in seizure cases. However, treatment should be conducted only under direct medical supervision. Consult a physician.

SHINGLES

Herpes Zoster or "shingles", a very painful skin rash, is caused by the chickenpox virus. After childhood chickenpox most of the viral organisms are destroyed but survivors can lie dormant in sensory nerves for many years. Later in life some people experience a decline in the efficiency of the immune system allowing the virus to reemerge and cause shingles. Shingles often follows a stressful episode. The area usually involved is a strip of skin over the ribs on one side or, less commonly a strip on one side of the neck and arm or the lower part of the body. Sometimes the upper half of the face on one side is affected.

When herpes zoster affects the chest/rib area, ceramic magnet(s) or 3-4 stacked magnetic strips should be applied over the rash area on the front of the body as well as on the backside of body at nerve endings.

If the rash is on the face ceramic magnet(s) or 3-4 stacked magnetic strips should be placed over the rash area a minimum of 3 hours daily or as frequently as needed for relief of symptoms.

The Magnetic Bed would provide optimum total body treatment for the viral infection. Also see Infections, Viral.

SHOPPING FATIGUE

The environmentally sensitive person often experiences symptoms during and after shopping. To prevent symptoms while shopping, wear a 3X4" plastiform magnetic strip on the crown of the head under a hat.

Ideal therapy after shopping exposure is 30-45 minutes of relaxation on a Magnetic Bed. If Magnetic Bed is not available, apply magnets as follows.

To treat exhaustion and other symptoms after shopping, try the Cerebellar Vestibular placement &/or place any size magnet(s) on the forehead for 15-30 minutes or until symptoms are relieved. The liver and spleen may also be treated directly to support the detoxification process.

Also refer to Chemical Exposures.

SHOULDER TENSION

Occupational, physical and emotional stress are often manifested in a tightening of the trapezius, a large muscle extending from the back of the skull to the lower part of the thoracic spine. To relieve tension place any size magnet(s) across the tender shoulder area and a 3X24" plastiform magnetic strip down the spine. In addition, wearing a magnetic wrist bracelet has been observed to be of value.

SINUSITIS

Inflammation of the membrane lining of facial sinuses (air-filled cavities in the bones surrounding the nose) is termed sinusitis. Symptoms can include tightness/fullness of affected area, a throbbing facial ache, stuffy nose, post nasal drip, etc. Various magnetic placements should be tried for best results, such as: Cerebellular Vestibular protocol or placement of any magnet(s) on the forehead. To achieve drainage of the sinuses, any of the magnets described herein may be applied to the side or below the sinuses to draw mucous out of the sinuses. Application may be as lengthy or as frequent as needed to relieve symptoms and discomfort.

SLEEP

Method I: Negative magnetic energy should be applied at the crown of the head nightly during sleep using 4 4X6X1/2" ceramic magnets standing upright l" apart. The crown of the head (top) should rest 2-3" from the magnets during sleep placing the head in the negative magnetic field all night.

During excitation (waking hours) the brain and central nervous system have a positive magnetic energy drive but during sleep the polarity of the brain and central nervous system swtiches to negative magnetic drive. The negative

magnetic field produced by the magnets at the crown of the head creates a natural setting for initiation and maintenance of deep sleep which restores biological energy. In addition, blood flowing through the brain in the magnetic field produces electromagnetic energy.

Method II: Secure a round Neodymium magnet on the crown of the head (exact center top of head). Not only will this enhance deep sleep with physical and mental energy but will treat the pineal gland responsible for production of melatonin. Melatonin in particular controls the cyclic pattern of sleep and wakefulness. Since the pineal gland is a magnetic organ it is sensitive to the daily cyclic patterns of the Earth's magnetic field and its function can be affected by the application of magnetic field exposure. Melatonin also governs biocycles which when disordered are stress producing and can relate to emotional, mental and physical illnesses.

SORE THROAT

For relief of sore throat, various magnetic placements should be tried for best effect. Magnetic therapy using ceramic magnet(s) or 3-4 stacked plastiform magnetic strips should be directly applied to the painful throat, 7th cervical and lymph nodes in the neck as described in Lymph Nodes & Carotid Arteries.

SPRAIN OR STRAIN

Ceramic magnet(s) or 3-4 stacked magnetic strips should be applied directly to the injured area to reduce inflammation and swelling.

STRESS, EMOTIONAL &/OR PHYSICAL

Ideally emotional or physical stress may be relieved by 30-60 minutes of relaxation on a Magnetic Bed which not only relieves physical tension but calms the central nervous system and soothes the emotional state.

If a Magnetic Bed is not available, a 3X24" plastiform magnetic strip should be applied to the spine while simultaneously applying the Cerebellar Vestibular protocol.

TENNIS ELBOW

For pain relief from "tennis elbow", place negative magnetic field on painful site holding in position with an Ace Bandage. It is extremely important not to allow the positive side of the magnet to come into contact with any other part of the body, such as the ribcage. Therefore, if the magnet is placed on the inside of the elbow it must be held away from the body during treatment. Leave magnet in place 30-60 minutes even if pain is relieved sooner. Treat a minimum of six times daily or as often as needed.

When tennis is resumed, treat elbow 30 minutes before play and 30 minutes after play for 6 weeks. Thereafter continue magnetic application 30 minutes daily for 4 weeks. If pain is present in more than one area, i.e., inside and outside elbow, two magnets may be used, one on each painful area. Alternatively, one magnet may be used on the inside of elbow for 30-60 minutes and then rotated to the outside for 30-60 minutes.

TENSION (General)

General stress and tension may be relieved by 30-60 minutes of relaxation on a Magnetic Bed which not only reduces physical tension but calms the central nervous system and soothes the emotional state.

If a Magnetic Bed is not available, a 3X24" plastiform magnetic strip should be applied to the spine while simultaneously applying the Cerebellar Vestibular protocol.

TINNITIS (Ringing in ears)

Food allergies should be considered as a cause for ringing in the ears. Another contributing factor can be metal fillings in the teeth setting up an electrical current in the mouth which can produce not only ringing in the ears but interference with the energy system. Magnetic therapy should be applied as described in Cerebellar Vestibular protocol. In addition, magnet(s) of any size may be tried directly over and around the ears. Use the placement which provides symptom relief. Treatment may be as lengthy and frequent as needed.

If ringing in the ears disappears when the magnetic field is applied and returns when the magnetic field is removed, a physician should be consulted regarding the possibility of electrical interference from metal fillings in the teeth.

TOOTHACHE

Ceramic magnet(s) or 3-4 stacked plastiform magnets should be applied directly to the painful area until symptoms are relieved. If infection is present, treatment should continue as continuously as possible until a dentist can be consulted. Magnetic energy reduces infection and inflammation and thus may relieve pain.

Late one Friday afternoon a male patient developed a throbbing toothache. Knowing it would be Monday before he could probably reach a dentist, he applied the negative pole of a magnet to the inflammed and swollen gum area which was obviously abcessed. Symptoms subsided and he enjoyed a relatively comfortable weekend. On Monday morning upon presenting himself at the dentist's office the tooth was x-rayed and declared to be free of infection. The inflammation did not return.

TOXICITY

The ideal treatment for toxicity would be full body therapy with a Magnetic Bed. If a Magnetic Bed is not available, Nightly Routine should be followed regularly with special attention given to the liver. The Nightly Routine may be used as frequently and as long as necessary (day as well as night) to relieve symptoms.

TRAUMA FROM FALLS, ACCIDENTS, ETC.

Magnetic therapy should be applied as quickly as possible to the injured area to reduce inflammation, swelling and pain. Any size magnets may be used--plastiform magnetic strips should be stacked for added strength.

A 79 year old woman sustained a hard blow to the forehead and a slight laceration of the skin from a fall against a wooden chest. A plastiform magnetic strip was immediately applied to the forehead and alternated every 30 minutes with an ice pack. As a result of this treatment the usual expected discoloration from such a blow was dramtically reduced and healing occurred quickly. No adverse symptoms developed.

A 12 year old boy fell and hit his head while playing basketball. He was able to ride his bicycle home to tell his mother that his head hurt and he was nauseated. The mother applied ceramic cube magnets to each temple and a 12" plastiform magnetic strip down the spine for 30 minutes. One hour later the boy had completely recovered and ran off to play and eat with his friends.

TRIGLYCERIDES Refer to Cholesterol & Triglycerides.

TUMORS

Benign tumors should be treated directly with ceramic magnet(s) or 3-4 stacked plastiform magnets. In addition a Magnetic Bed or the Nightly Routine should be followed regularly. A physician should be consulted.

ULCERS Refer to Gastrointestinal Disorders.

URINARY PROBLEMS

The kidneys should be treated directly with ceramic magnet(s) or 2-4 stacked

plastiform magnets for 30-60 minutes twice daily or more if needed. The kidneys are in the back on the right and left side of the spine below the rib cage.

For additional treatment information, refer to Infections, Bacterial and Infections, Fungal.

VAGINAL INFECTION Refer to Infection, Fungal.

WARTS, MOLES, SKIN LESIONS
Warts and moles may be effectively treated by placing any size magnet(s), preferrably a round neodymium, directly on the skin area involved for several hours daily or all night. Magnets may be bound in place with Ace Bandage or taped on the affected site.

WEIGHT REDUCTION
Method #1: Magnetic Bed
l) Sleep nightly on Magnetic Bed
2) During waking hours place ceramic magnet(s) or 3-4 stacked plastiform magnets on local areas of fatty tissue for "spot reduction".
3) To suppress appetite place ceramic magnet(s) over the epigastric area 30 minutes premeal and during meals.

NOTE: It is advised that magnets NOT be applied directly over the abdomen for 60-90 minutes after a meal allowing sufficient time for peristalsis to occur.

Method #2: Solid State Magnets
Negative magnetic field energy is alkalinizing and, therefore, theoretically will dissolve fatty tissue deposits.
l) Direct application of negative magnetic energy to the fatty tissues (abdomen, buttocks, thighs, arms, etc.) is required using ceramic magnet(s) or 3-4 stacked plastiform strips. Since there is a direct relationship between the length of exposure to the magnets and the degree of fat resolution, the mid to low abdomen is best treated while asleep.
2) During waking hours, fatty deposits should be treated 30 minutes before each meal.
3) For appetite supressant, place ceramic magnet(s) over the epigastric area 30 minutes premeal and during meals.

NOTE: It is advised that magnets NOT be applied directly over the abdomen for 60-90 minutes after a meal allowing sufficient time for peristalsis to occur.

Magnets can be held in place by Ace Bandage, abdominal binders, pantyhose, etc.

It must be understood that magnetic dissolution of fat as described in Method

#2 is a steady but slow process. A Wyoming patient writes, "I have been using the magnets every day for allergies and have put one on my tummy at night. Have lost all the tummy fat and that which surrounds the waist. It really does work but takes a while. Guess it (the fat) takes a while to get there in the first place. Have felt so much better since using the magnets. . . ."

Although it is possible to reduce weight with negative magnetic energy without change in diet or physical routine, it is not wise to rely solely on this type of weight reduction for optimum health. Addictive states, which often play a primary role in obesity, have a degenerative effect on the biological system. Addictions may involve foods, caffeine, alcohol or nicotine. Addictions can produce l) an uncontrollable urge to eat for symptom relief, 2) impulsive food binging or overeating as well as 3) irresistible cravings--often for allergenic substances. No matter whether overeating is a result of addiction or habit, it is involved in obesity.

Logically a reduction in caloric intake should be considered.
Since food allergies and/or sensitivities behave much the same as addictions (with an initial withdrawal phase), it is wise to sort out food allergies/sensitivities by meals of single foods after eating infrequent foods (foods not normally eaten) for five days (Refer to <u>Brain Allergies: the PsychoNutrient Connection</u> by W.H. Philpott, M.D.). Temporary removal of allergenic and/or addictive foods from the diet can materially reduce food cravings and the urge to overeat.

<u>WHIPLASH</u>
Attach a plastiform magnetic strip of any length to the neck at the collar holding in place with tape or binder. Magnet may be worn all day or night as needed for pain relief.

<u>WITHDRAWAL SYMPTOMS</u> Refer to Addictive States.

THE LYMPHATIC SYSTEM

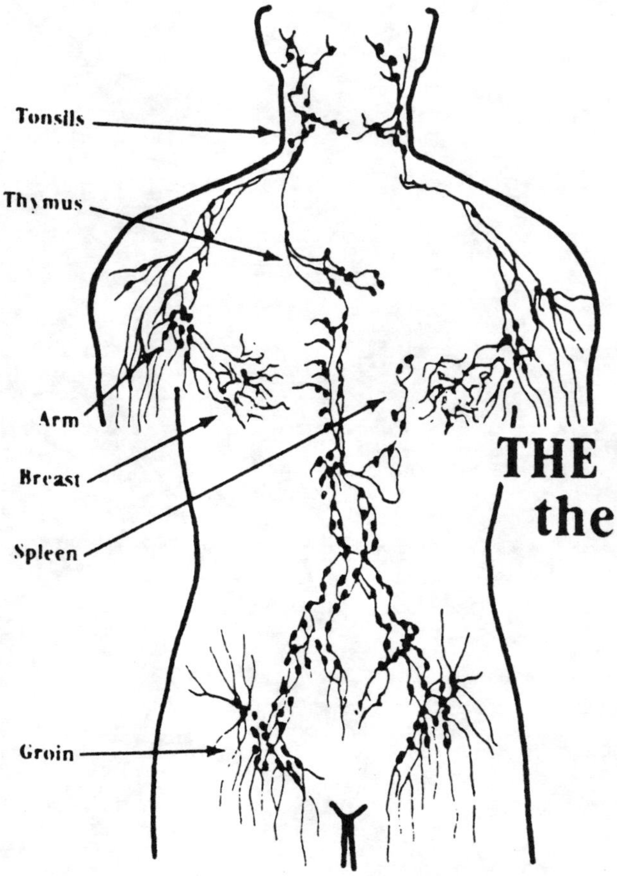

THE ENDOCRINE GLANDS . . . the remote control organs

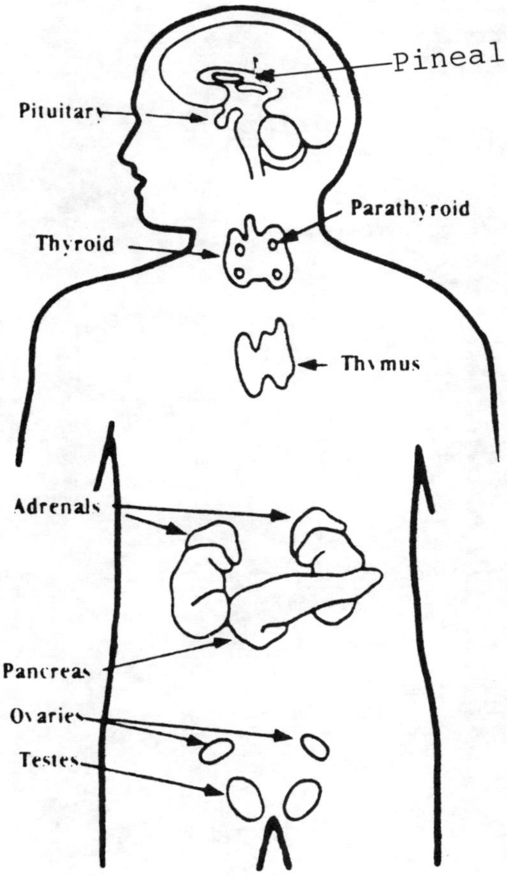

THE RESPIRATORY SYSTEM

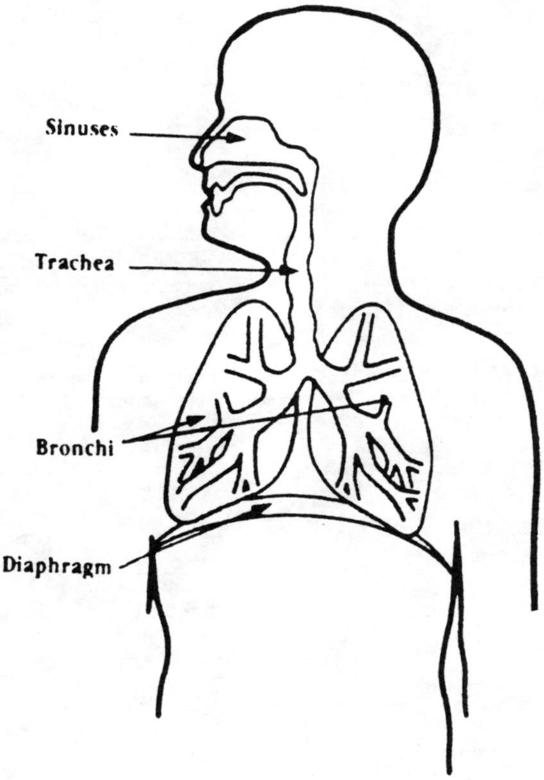

THE DIGESTIVE SYSTEM

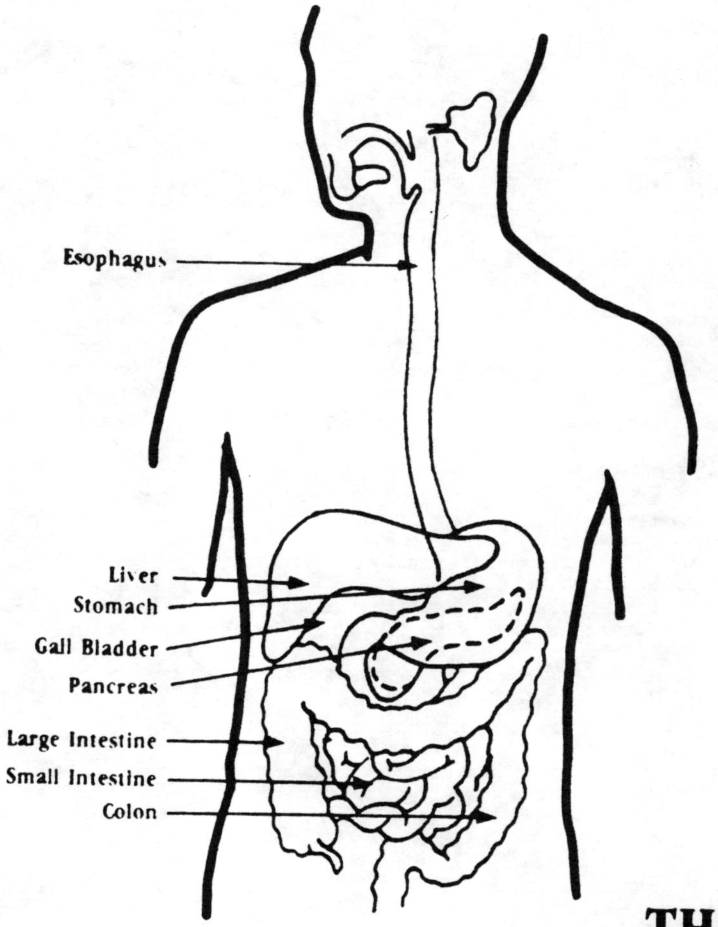

THE URINARY SYSTEM

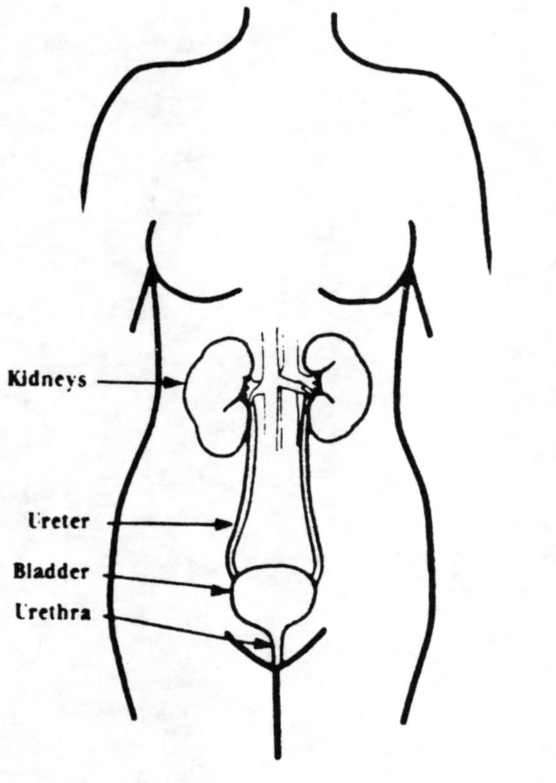

Positive Magnetic Energy Therapeutic Application

There are several valuable and safe applications of positive magnetic energy which should only be applied under medical supervision.

Negative magnetic energy should always be used in conjunction with positive magnetic energy to counter the potential for producing addiction, seizures, microroganism and parasite activity, inflammation, oxygen deficit of tissues, etc. Therefore, I will not describe here the therapeutic uses of positive magnetic energy. The Reader desiring further information in this area should contact a physician skilled in the application of magnetic therapy.

Additionally, positive magnetic energy has a useful place in facilitating responses for test purposes.

References Sources

BECKER, RO Cross Currents. Jeremy P. Tarcher, Inc.: Los Angeles, CA, l990

BECKER, RO & MARINO, AA. Electromagnetism & Life. State University of New York Press: Albany, NY l982

BECKER, RO & SELDEN, G. The Body Electric: Electromagnetism & the Foundation of Life. William Morrow & Co: New York, l985

BRODEUR, PAUL. Currents of Death: Power Lines, Computer Terminals, & the Attempt to Cover Up their Threat to Your Health.
Simon & Schuster Publishers: New York, 1989

CALLAHAN, PHILIP. Ancient Mysteries, Modern Visions: the Magnetic Life of Agriculture. Acres USA: Kansas City, MO, l984

DAVIS, AR & RAWLS, W. The Magnetic Blueprint of Life. Acres USA; Kansas City, MO, l979

DAVIS, AR & RAWLS, W. The Magnetic Effect. Acres USA: Kansas City, MO, l975

DAVIS, AR. The Anatomy of Biomagnetism. Acres USA: Kansas City, MO, l975

Encyclopedia Britannica, 15th Ed, Vol 18, Pg. 274, l986

FONTENAT JOHN, MD & LEVINE, STEPHEN PhD. "Melatonin Deficiency: Its Role in Oncogensis & Age-related Pathology". Journal of Orthomolecular Medicine, 5:1, l990

HALLETT, MARK, MD & COHEN, LEONARDO, MD. Journal of American Medical Assn., "Magnetism: A New Method for Stimulation of Nerve & Brain". Vol 262:4, 7/28/89

LEE, EW. Magnetism: An Introductory Survey. Daver Publications, Inc., NY, NY. l970

NAKGAWA, KYOICHI, MD. "Magnetic Field Deficiency Syndrome & Magnetic Treatment", Japan Medical Journal, #2745, 12/4/76

NUNEZ, PAUL. Electric Fields of the Brain: The Neurophysics of EEG. Oxford University Press: Oxford, NY, l98l

PHILPOTT, WH & KALITA, D. Brain Allergies: the PsychoNutrient Connection. Keats Publishing, New Canaan, CT, l981

PHILPOTT, WH & KALITA, D. Victory Over Diabetes: a BioEcologic Triumph. Keats Publishing, New Canaan, CT, l983

REYNOLDS, GAVIN. "Increased Concentrations & Lateral Asymmetry of Amygdala Dopamine in Schizophrenia", Nature. Cambridge, England. 305: 527-529, October 6, l983

STEVENS, JANICE, M.D., "EEG in Schizophrenia", CME Syllabus. APA 139th Annual Meeting, l986

USDIN, EARL, Ed. Endorphins in Mental Health Research. Oxford University Press: New York, l979

WATSON, STANLEY & MADDEN, JOHN IV. (USDIN, Earl, Hamburg, David, BARCHAS, Jack Edtrs). "Melatonin & Other Pineal Substances: Psychiatric & Neurological Implications". Neuroregulations & Psychiatric Disorders, Oxford University Press, NY, NY, l977

WEINBERGER, DANIEL, M.D. & KLEINMAN, JOEL, M.D. "Observations on the Brain in Schizophrenia", APA Annual Review, #5. American Psychiatric Press: Washington DC, l986

GLOSSARY

Anterior - situated before or toward the front

Antioxidant - a substance that opposes oxidation or inhibits reactions promoted by oxygen or peroxides

Arrhythmia - a lack of rhythm; alteration in rhythm of the heartbeat either in time or force

Bifurcation - the point at which an artery divides into two branches; becomes two pronged; forks

Biochemistry - chemistry that deals with chemical reactions and processes in living organisms

Biomagnetics - natural science that deals with application of magnetic fields to living organisms

Biophysics - branch of science concerned with the application of physical prinicples and methods to biolgoical problems

Bipolar - having or involving the use of two magnetic poles (north and south pole)

Bitemporal - relating to both temporals or temporal bones

Bradycardia - relatively slow heartbeat whether physiological or pathological

Cell mutation - a significant and basic alteration at cellular level

DNA - any of various nucleic acids that are localized especially in cell nuceli, are the molecular basis of heredity in many organisms

Edema - swelling

Electrolysis - the producing of chemical changes by passage of an electric current through an electrolyte

Electromagnetic contamination - presence of invisible vibrational frequencies and magnetic fields which may affect biological systems

Endorphins - any of a group of proteins with potent analgesic properties that

occur naturally in the brain; a self made narcotic

Extracellular - outside of the cells

Gauss - a unit of magnetic field intensity

Geomagnetic - of or relating to terrestrial magnetism

Hypoxia - decreased amount of oxygen in organs and tissues

Hypersensitivity - abnormal sensitivity

Intercellular - between or among cells

Magnet - a body which has the property of attracting particles of iron, cobalt, nickel or any of various metallic alloys and which has magnetic polarity

Maladaptive reaction - an nonimmunologic reaction to a substance

Motility - the power of spontaneous movement

Occiput - back of the head

Oxidation - to combine with oxygen

Oxygenation - to impregnate, combine or supply (as blood) with oxygen

Paramagnetic - having the property of paramagnetism (the property of being magnetic as shown by assuming a position parallel with the lines of force between the two poles of a magnet)

Peristalsis - muscular wavelike movement of the intestinal tract

Posterior - situated behind; as the hinder parts of the body

Tachycardia - relatively rapid heart beat whether physiological as after exercise or pathological

Unipolar - having but one magnetic pole

William H. Philpott, M.D., has enjoyed the practice of medicine for 40 years. He has specialized in offering both physically and mentally ill patients scientific advances in medicine which provided answers for otherwise unsolved medical problems. His books, Brain Allergies: the PsychoNutrient Connection, Victory Over Diabetes: a BioEcologic Triumph, New Hope for Physical & Emotional Illnesses and The BioMagnetic Handbook reveal the progressive movements occurring in medicine in general and in his practice in particular during his medical career.

Dr. Philpott is devoting himself to conducting self-help seminars reflecting his belief that when properly informed people can help themselves achieve and maintain health and happiness. A major goal of his writings is to educate the public not only in self-help but in how to assist their personal physician in finding answers to health problems.

BioMagnetic Handbook highlights the values of the rapidly expanding scientific field of electromagnetics in the practice of medicine. This new movement in energy medicine sharpens the tools of physicians and broadens the scope of self-help health restoring and health maintaining measures.

Dr. Philpott serves as staff consultant to American Biologics Hospital-Mexico and is also involved in research advancement in biomagnetics.